Landscapes in the Eastern Mediterranean between the Future and the Past

Landscapes in the Eastern Mediterranean between the Future and the Past

Special Issue Editors

Ioannis N. Vogiatzakis
Theano S. Terkenli
Maria Gabriella Trovato
Nizar Abu-Jaber

MDPI • Basel • Beijing • Wuhan • Barcelona • Belgrade

Special Issue Editors
Ioannis N. Vogiatzakis
Open University of Cyprus
Cyprus

Theano S. Terkenli
University of the Aegean
Greece

Maria Gabriella Trovato
American University of Beirut
Lebanon

Nizar Abu-Jaber
German Jordanian University
Jordan

Editorial Office
MDPI
St. Alban-Anlage 66
4052 Basel, Switzerland

This is a reprint of articles from the Special Issue published online in the open access journal *Land* (ISSN 2073-445X) from 2017 to 2018 (available at: https://www.mdpi.com/journal/land/special_issues/mediterraneanlandscapes)

For citation purposes, cite each article independently as indicated on the article page online and as indicated below:

LastName, A.A.; LastName, B.B.; LastName, C.C. Article Title. *Journal Name* **Year**, *Article Number*, Page Range.

ISBN 978-3-03921-774-8 (Pbk)
ISBN 978-3-03921-775-5 (PDF)

Cover image courtesy of Dr Paraskevi Manolaki.

Contents

About the Special Issue Editors

Ioannis N. Vogiatzakis is Professor of Environmental Conservation at the Open University of Cyprus. His research interests include ecology and biogeography of Mediterranean islands and mountains; the links between geomorphology and vegetation distribution patterns; predictive vegetation and habitat mapping; landscape-based approaches to nature conservation delivery; effectiveness of protected areas for biodiversity conservation; effects of landscape structure and habitat quality on biodiversity; spatial scale effects on ecosystem services; and sustainability assessment. He has published extensively and has edited two books on Mediterranean Island Landscapes, 2008 and Mediterranean Mountain Environments, 2012.

Theano S. Terkenli is Professor and founding member of the first Department of Geography in Greece since 1994 and faculty and board member of the Interdepartmental Graduate Program in 'Strategic Management of Tourist Destinations and Hospitality Businesses' at the University of the Aegean since 1999. Her Ph.D. in Geography was awarded by the University of Minnesota, Twin Cities; M.Sc. in Landscape Architecture by University of Wisconsin, Madison; and B.Sc. in Forestry and Environmental Science by the Aristotle University of Thessaloniki. Terkenli is an active member, conference organizer, and/or board member of several international and Greek associations and organizations. Her research/publishing/teaching specialty and interests encompass the fields of Cultural Geography, Landscape Geography, and Critical Perspectives to Tourism. She is the author and/or editor of a number of books, including Connections, Mobilities, Urban Prospects and Environmental Threats: The Mediterranean in Transition, 2015; Human Geography: Humans, Society and Space, 2007; Landscapes of a New Cultural Economy of Space, 2016; Tourism Geography, 2003; The Cultural Landscape: Geographical Perspectives, 1996.

Maria Gabriella Trovato is Assistant Professor of Landscape Architecture at the Landscape Design and Ecosystem Management System at the American University of Beirut. Her research focus is summarized into the following major areas: 1. Landscape and social/ environmental justice; 2. Landscape in emergencies and new challenges (climate change, depletion of natural resources, conflict between globalization and local development, and re-localization of war refugees); 3. Landscape mapping and innovation in planning.

Nizar Abu-Jaber, Professor of Geology, is interested in the relationships between landscape, water resources, and management of landscape that ensure the long-term sustainability of surface and groundwater. This includes an understanding of water dynamics related to recharge and subsurface movement, as well as the maintenance of water quality under human pressures.

Preface to "Landscapes in the Eastern Mediterranean between the Future and the Past"

Landscape means different things to different people, both literally and metaphorically. It results from the interplay of physical, natural, and human/cultural elements of our surroundings and the way that people perceive these interactions. Landscape character is not just about the elements that make up a landscape but also embraces the aesthetics and perceptual factors that make different places distinctive. Different combinations of these elements create the distinctive character of landscapes in places, allowing different landscapes to be mapped, analyzed, and described (Vogiatzakis, 2011). Landscapes have thus long been viewed as 'multifunctional', integrating ecological, economic, sociocultural, historical, and aesthetic dimensions. This is also reflected in the European Landscape Convention as well as in the related methodologies developed for landscape conservation, and management (Vogiatzakis, 2011).

Landscape is considered one of the key themes of policies for environmental and territorial sustainability and has been on the political agenda of European countries, resulting in innovations in Land and Spatial Policies and in specific sectors such as agriculture and cultural heritage (Peano and Cassatella, 2011). Being an essential component of people's surroundings, landscape is an expression of the diversity of their shared cultural and natural heritage and the foundation of their identity. Therefore, besides being a source of local pride and quality of life, landscapes are also key to the development of tourism or to drawing other social, cultural, or economic investment to a specific region or place (Terkenli, 2014). They constitute a critical environmental/economic/social/cultural resource for implementing sustainable development (Pavlis & Terkenli, 2017). Landscape-scale approaches are fundamental to understand the past and present of a landscape's constituent elements, and are now considered to be an appropriate spatial framework for the analysis of sustainability (Morse et al., 2011).

At the basis of landscape definition and articulation lie human ways of interrelating with the land and landed resources. These ways differ and vary around the world, but have increasingly been obliterated by the Western scenic landscape paradigm, at the basis of most Western design and planning disciplines and professions involved in landscape stewardship. Nonetheless, globalizing contemporary needs and dictates requires common strategies, models, aims, and methods, in order to address landscape-related problems and prospects around the world. This volume is the result—a response towards this goal which aims to facilitate the broader application of landscape character assessment in an Eastern Mediterranean context, in accordance with local and regional conditions and particularities, by critically extrapolating methods and techniques which are well-established in Europe.

Ioannis N. Vogiatzakis, Theano S. Terkenli, Maria Gabriella Trovato, Nizar Abu-Jaber
Special Issue Editors

References

Morse, S.; Vogiatzakis, I.N.; Griffiths, G.H. Space and sustainability. Potential for landscape as a spatial unit for assessing sustainability. *Sustain. Dev.* **2011**, *19*, 30–48.

Pavlis, E.; Terkenli, T.S. Landscape values and the question of 'cultural sustainability': Exploring an uncomfortable relationship in the case of Greece. *Nor. J. Geogr.* **2017**, *3*, 168–188.

Peano, A.; Cassatella, C. Landscape assessment and monitoring. In *Landscape Indicators*; Springer: Dordrecht, The Netherlands, 2011; pp. 1–14.

Terkenli, T.S. Landscapes of Tourism. In *The Wiley-Blackwell Companion to Tourism*, 1st ed.; Lew, A.A., Hall, M.C., Williams, A.M., Eds.; John Wiley & Sons, Ltd.: Oxford, UK, 2014; pp. 282–293.

Vogiatzakis I.N. Mediterranean experience and practice in Landscape Character Assessment. *Ecol. Medit.* **2011**, *37*, 17–31.

land

MDPI

Editorial

Landscapes in the Eastern Mediterranean between the Future and the Past

Ioannis N. Vogiatzakis [1,*], Theano S. Terkenli [2], Maria Gabriella Trovato [3] and Nizar Abu-Jaber [4]

1 School of Pure & Applied Sciences, Open University of Cyprus, PO Box 12794, 2252 Nicosia, Cyprus
2 Department of Geography, University of the Aegean, 81100 Mytilini Lesvos, Greece; t.terkenli@aegean.gr
3 Landscape Design and Ecosystem Management, Faculty of Agricultural and Food Sciences, American University of Beirut, Riad El-Solh, Beirut 1107-2020, Lebanon; mt63@aub.edu.lb
4 Center for the Study of Natural and Cultural Heritage, German Jordanian University, P.O. Box 35247, Amman 11180, Jordan; nizar.abujaber@gju.edu.jo
* Correspondence: ioannis.vogiatzakis@ouc.ac.cy, Fax: +357 22411601

Received: 7 December 2018; Accepted: 13 December 2018; Published: 18 December 2018

Landscapes have long been viewed as complex, synthetic entities reflecting the human imprint upon the land. Efforts have been made for a number of decades now to plan, manage, and use them on a "multi-functional" and sustainable basis, integrating all of their dimensions—ecological, economic, socio-cultural, historical, aesthetic, etc. (Wascher 2000 [1]; Wu 2013 [2]). Landscape scale approaches have been fundamental to the understanding of past and present cultural evolution and are now considered to be an appropriate spatial framework for the analysis of land systems sustainability (Buttimer 1998 [3]; Terkenli 2005 [4]). The methods and tools of landscape analysis and intervention have also greatly evolved since their early development in Europe and North America (Fairclough et al. 2018 [5]). Although significant progress has been made, there remain many issues which are understudied or not investigated at all, at least in a Mediterranean context, as local contexts greatly impact the applicability of landscape characterization and planning approaches.

This special volume in LAND focuses on the application of landscape theory and practice in the Eastern Mediterranean and mainly reports on the outcomes of an international conference held in Jordan in December 2015 upon the closure of the MEDSCAPES PROJECT (Vogiatzakis et al. 2015 [6]), with the title "International Conference on Landscapes of the Eastern Mediterranean: challenges, opportunities, prospects and accomplishments". The focus of this volume constitutes a timely and growing area of research interest, not only because these landscapes have so far been understudied, but also because the Eastern Mediterranean region is a rich site of strikingly variegated, long-standing, multi-cultural human-environmental interactions (Braudel 1972 [7]; King et al. 1997 [8]; Grove and Rackham 2001 [9]; Lowenthal 2015 [10]). These interactions, resting on and taking shape through millennia of continuity in tradition, have been striving to adapt to technological advances, while currently juggling numerous multi-layered socio-economic and climate-environmental crises.

Landscape science and public awareness in Europe have been progressing in leaps and bounds. The challenges involved in landscape-related issues and fields are, nonetheless, still multiple, and refer to landscape stewardship and protection as well as to the development of comprehensive relevant theoretical and methodological approaches, in tandem with public sensitization and participatory governance, and in coordination with appropriate top-down planning and policy instruments. By definition (Council of Europe, 2000 [11]), landscape has long been established as the medium and product of human ways of relating to our surroundings. Significant inroads have been made in recent research endeavors towards almost all aspects of landscape analysis and interventions, including landscape ecology, rural sociology, and urban gentrification. For instance, the theory of landscape ecology, founded and described in the seminal work by Forman and Godron (1986) [12], set the scene for a paradigm shift in ecology from site-based to landscape-based processes. Although the landscape may appear to be a principally material or physical entity, it is experienced as a synthetic whole through living, traveling, or simply seeing it from afar, as illustrated by recent scientific perspectives, such as "interactive", "enactive", "embodied", or "more-than-representational" geographies. The erroneous distinction between natural and cultural landscapes is still encountered: an artificial distinction, since, in the Mediterranean—and particularly the Eastern Mediterranean—very few areas free of human intervention remain (Vogiatzakis 2012 [13]; Vogiatzakis et al. 2008 [14]). Indeed, Mediterranean landscapes merit increased and more focused attention, on the basis of their historically continuous and heavy exposure to human impact (Lowenthal 2015 [10]), exacerbated, in recent decades, by conditions of strife and crisis.

The significance, then, of this task of applying conventional Western knowledge and practices to this part of the world first lies in filling a gap. The Eastern Mediterranean has been relatively lagging and in great need of such methods and spatial interventions vis-à-vis its landscapes, but finding itself increasingly at a crossroads where serious such efforts are starting to be undertaken. Meanwhile, it represents an especially fragile part of the earth, where cultural and environmental resource deterioration has been progressing at rapid rates (i.e. desertification, water depletion, cultural destruction, ecological impoverishment etc.). On the basis of the fragility of the Eastern Mediterranean region, due to its location, fragmented and confined configuration, intricate geomorphology and tectonism, but also due to its vulnerability (i.e. to human use, through tourism, poor or lacking planning practices, overpopulation and heavy urbanization), this collective effort becomes quite timely. It is timely not only because of rampant environmental change, presenting a series of risks and hazards, especially when combined with detrimental human impact on the land; it becomes timely and useful also because of growing forces of change and recent socio-political upheaval in this part of the world (wars, socio-economic crises, geopolitical tensions). These trends call for new inroads in landscape risk assessment methodologies, in the context of landscape decision-making and support systems. Furthermore, new information and communication technologies, referring to and assisting almost all sectors of human activity, present ever-expanding arenas of human-landscape interactions. All of these methods, tools, and technologies are showing signs of being enthusiastically embraced by the societies of the region here under focus.

While the transfer of Western knowledge and practices to the Middle East has been more or less established, the MEDSCAPES project offered some interesting reverse possibilities and potential. In this regard, a promising example concerns public participation in landscape stewardship, through lessons acquired from the Arab-speaking world, namely the HIMA traditions. Participatory practices in landscape planning and management that we have begun to apply in the 'Western' world during the past few decades have been commonplace custom in the 'Near-Eastern' world for millennia.

Finally, this undertaking highlights an area begging for further scientific exploration in terms of the long-standing historically-rooted connections and interactions between the peoples of the Eastern Mediterranean region, and continuously bearing fruit in terms of revealing its closely-knit common cultural character and trajectories. Just a few examples: Greece was enriched by Egypt and Phoenicia, Rome by Greece, Byzantium by the Balkans, Southeastern Europe by the Middle East, and so on. Most significantly, though, this collective task becomes very relevant because it attempts to elucidate and understand the very different ways in which indigenous populations of the Eastern Mediterranean have related to their surroundings through time and space, as well as their contemporary rich and variegated traditions and concepts of landscape and space.

This is broadly reflected also in the structure of this volume. The volume begins with the laying out of the cultural-historical context and significance of the project, proceeds to an analysis of Landscape Character Assessment (LCA) methodological applications and their outcomes, and closes with planning practices and tools as well as empirical lessons drawn from their implementation. In so doing, it places the discussion of specific landscapes in their socio-political and environmental contexts, applies geospatial analysis for landscape management but also reveals the role of people (stakeholders) and societies in LCA and landscape planning. More specifically, Moraitis [15] sets the scene for tracing the roots to common traditions in the Arabic and Byzantine cultures; these cultures seem to have played a crucial role in the subsequent shaping of Eastern Mediterranean landscapes. In Cyprus, sacred landscapes of the Archaic and Hellenistic periods are accordingly discussed by Papantoniou and Vrionis [16] as constructs of religion, politics, identity, and memory, whose change through time has been a result of socio-political and economic transformation and adaptation to changing contexts and circumstances.

Transferring methodologies and concepts from one geographical context to another is not without challenges. LCA is such an example (Fairclough et al. 2018 [5]). LCA has a significant contribution to make as a spatial framework for the rekindled concept of "multi-functional landscapes", referring to landscape provision of a range of functions, services, and human-derived benefits. Although the application of LCA in the Mediterranean region is on the increase (Vogiatzakis 2011 [17]), the paper by Griffiths [18] is perhaps the first of its kind to review the implementation and spread of LCA in the Eastern Mediterranean, effectively starting in Cyprus, before it was embraced by MEDSCAPES for a number of other case studies. Notwithstanding what may often be perceived as a uniform area (culturally, historically, physiographically), there are distinct differences between and within countries of the Eastern Mediterranean, as demonstrated in today's landscapes. Therefore, it becomes difficult to make meaningful comparisons between areal units of analysis without reference to a common spatial framework. In their paper, Vogiatzakis and Manolaki [19] use such a framework to investigate the diversity and variability of mapped landscapes, in six case studies across four Eastern Mediterranean countries.

Socio-political processes, followed by ecological system responses in time, imprint their impacts on present-day landscapes. Not only agricultural intensification and land use abandonment, but also mass tourism and migration, constantly shape new landscapes in the Mediterranean. These changes have had an impact on resources and biodiversity—some demonstrated imminently, while others, like biodiversity loss, with a time lag effect (Lindborg and Eriksson 2004 [20]). Geospatial analysis has brought new perspectives to landscape studies, by allowing variable reconstructions of the past, and novel ways of representing the present and predicting the future (Vogiatzakis and Melis 2015 [6]). As exemplified herein by a case study from southern Italy (Statuto et al [21], this volume), the developed methodology can be used to provide insights into present landscape monitoring and management, by weaving changes of the past into decision-making processes.

Landscape meanings and messages are often within the eye of the beholder, a challenge that has accompanied all landscape-related disciplines from the start. Landscape users and stakeholders come from a variety of groups with different views and often conflicting interests, in the use of landscape as space and as resource. Stakeholders' involvement in landscape assessment is of utmost importance and also advocated by the European Landscape Convention (Council of Europe 2000 [11]). The paper by Terkenli and Kavroudakis [22] in this issue assesses high-level landscape stakeholders' perceptions and understandings of landscape-related issues, threats and problems in the Eastern Mediterranean through a purposive comparative research survey of four case studies: Cyprus, Greece, Jordan and Lebanon. The paper by Makhzoumi and Al-Sabbagh [23] further illustrates this perspective, by providing an example of stakeholders' engagement in participatory planning processes in Lebanon.

Finally, planning applications of LCA have been widely employed in Europe over the last twenty years (Marušič and Jančič 1998 [24]; Wascher and Jongman 2003 [25]; Mücher et al. 2010 [26]), but are still used to a limited degree in the Mediterranean (Loupa-Ramos and Pinto-Correia 2018 [27]). In this issue, Odeh et al. [28] apply landscape characterization to two case studies in Jordan and demonstrate ways of including the results into current land-use planning practices in the country. In addition, Trovato et al. [29] provide a landscape risk assessment method and landscape decision support system (LDSS), conducted on the regional scale of the Eastern Mediterranean basin. The study is an important attempt to assess and collate the landscape character of the region by using comparable indicators to evaluate the risk of loss of landscape value in the Eastern Mediterranean. Furthermore, the LDSS enables users and stakeholders to build helpful scenarios in the planning process in selected areas.

Nonetheless, this volume sets out to fulfill its goals, subject to a series of delimitations, which only serve to open the ground and create space for much further and more in-depth future scientific exploration of Eastern Mediterranean landscapes, as well as ways, methods and tools to plan, manage and study them.

Author Contributions: The editorial was conceptualized by T.S. Terkenli and I.N. Vogiatzakis, M-G Trovato and N. Abu-Jaber reviewed and edited the text.

Funding: This work was partly funded by the MEDSCAPES project (ENPI-CBCMED program).

Acknowledgments: We would like to thank the German Jordanian University for organizing and hosting the International Conference on Landscapes of the Eastern Mediterranean.

Conflicts of Interest: The authors declare no conflict of interest.

References

1. Wascher, D.M. (Ed.) *Landscapes and Sustainability: Proceedings of the European Workshop on Landscape Assessment as a Policy Tool, 25–26 March 1999, Strasbourg, France*; European Centre for Nature Conservation: Tilburg, The Netherlands; European Centre for Nature Conservation and The Countryside Agency: Brandon Marsh, UK, 2000.
2. Wu, J. Landscape sustainability science: Ecosystem services and human well-being in changing landscapes. *Landsc. Ecol.* **2013**, *28*, 999–1023. [CrossRef]
3. Buttimer, A. Landscape and life: Appropriate scales for sustainable development. *Irish Geogr.* **1998**, *31*, 1–33. [CrossRef]
4. Terkenli, T.S. New landscape spatialities: The changing scales of function and symbolism. *Landsc. Urban Plan.* **2005**, *70*, 165–176. [CrossRef]
5. Fairclough, G.; Herlin, I.S.; Swanwick, C. *Routledge Handbook of Landscape Character Assessment*; Routledge: London, UK, 2018.
6. Vogiatzakis, I.N.; Manolaki, P.; Trigkas, V. (Eds.) LCA training and Implementation. In *Medscapes WP5 Final Report*; Open University of Cyprus: Nicosia, Cyprus, 2016.
7. Braudel, F. *The Mediterranean and the Mediterranean World in the Age of Phillip II, Vol 1*; HarperCollins Publishers: New York, NY, USA, 1972.
8. King, R.; Proudfoot, L.; Smith, B. (Eds.) *The Mediterranean: Environment and Society*; Taylor & Francis: London/Arnold, UK, 1997.
9. Grove, A.T.; Rackham, O. *The Nature of Mediterranean Europe: An Ecological History*; Yale University Press: New Haven, CT, USA, 2001.
10. Lowenthal, D. Geography, history and heritage: A Mediterranean overview. In *Connections, Mobilities, Urban Prospects and Environmental Threats: The Mediterranean in Transition*; Terkenli, T., Douguédroit, A., Cassar, L.F., Eds.; Cambridge Scholars Publishing: Newcastle upon Tyne, UK, 2015; pp. 1–50.
11. Council of Europe. *European Landscape Convention*; European Treaty Series—No. 176; Council of Europe: Florence, Italy, 2000.
12. Forman, R.; Godron, M. *Landscape Ecology*; John Wiley: New York, NY, USA, 1986.
13. Vogiatzakis, I.N. (Ed.) *Mediterranean Mountain Environments*; Wiley-Blackwell: Oxford, UK, 2012.
14. Vogiatzakis, I.N.; Pungetti, G.; Mannion, A. (Eds.) *Mediterranean Island Landscapes: Natural and Cultural Approaches*; Landscape Series Volume 9; Springer Publishing: New York, NY, USA, 2008.
15. Moraitis, K. The Cultural Landscape Past of the Eastern Mediterranean: The Border Lord's Gardens and the Common Landscape Tradition of the Arabic and Byzantine Culture. *Land* **2018**, *7*, 28. [CrossRef]
16. Papantoniou, G.; Vionis, A.K. Landscape Archaeology and Sacred Space in the Eastern Mediterranean: A Glimpse from Cyprus. *Land* **2017**, *6*, 40. [CrossRef]
17. Vogiatzakis, I.N. Mediterranean experience and practice in Landscape Character Assessment. *Ecol. Mediterr.* **2011**, *37*, 17–31.
18. Griffiths, G. Transferring landscape character assessment from the UK to the Eastern Mediterranean: Challenges and perspectives. *Land* **2018**, *7*, 36. [CrossRef]
19. Vogiatzakis, I.N.; Manolaki, P. Investigating the diversity and variability of Eastern Mediterranean Landscapes. *Land* **2017**, *6*, 71. [CrossRef]

20. Lindborg, R.; Eriksson, O. Historical landscape connectivity affects present plant species diversity. *Ecology* **2004**, *85*, 1840–1845. [CrossRef]
21. Statuto, D.; Cillis, G.; Picuno, P. Using Historical Maps within a GIS to Analyze Two Centuries of Rural Landscape Changes in Southern Italy. *Land* **2017**, *6*, 65. [CrossRef]
22. Terkenli, T.S.; Kavroudakis, D. A Stakeholders' Analysis of Eastern Mediterranean Landscapes: Contextualities, Commonalities and Concerns. *Land* **2017**, *6*, 90. [CrossRef]
23. Makhzoumi, J.; Al-Sabbagh, S. Landscape and Urban Governance: Participatory Planning of the Public Realm in Saida, Lebanon. *Land* **2018**, *7*, 48. [CrossRef]
24. Marušič, J.; Jančič, M. *Regional Distribution of Landscape Types in Slovenia: Methodological Bases*; Ministry of Environment and Physical Planning: Ljubljana, Slovenia, 1998.
25. Washer, D.; Jongman, R. (Eds.) *European Landscapes: Classification, Evaluation and Conservation*; Environment Technical Reports; European Environment Agency: Copenhagen, Denmark, 2003.
26. Mücher, C.A.; Klijn, J.A.; Wascher, D.M.; Schaminée, J.H.J. A new European Landscape Classification (LANMAP): A transparent, flexible and user-oriented methodology to distinguish landscapes. *Ecol. Indic.* **2010**, *10*, 87–103. [CrossRef]
27. Loupa-Ramos, I.; Pinto-Correia, T. Landscape Character Assessment across scales: Insights from the Portuguese experience of policy and planning. In *Routledge Handbook of Landscape Character Assessment*; Routledge: London, UK, 2018; pp. 124–135.
28. Odeh, T.; Boulad, N.; Abed, O.; Abu Yahya, A.; Khries, N.; Abu-Jaber, N. The Influence of Geology on Landscape Typology in Jordan: Theoretical Understanding and Planning Implications. *Land* **2017**, *6*, 51. [CrossRef]
29. Trovato, M.G.; Ali, D.; Nicolas, J.; El Halabi, A.; Meouche, S. Landscape risk assessment model and decision support system for the protection of the natural and cultural heritage in the Eastern Mediterranean Area. *Land* **2017**, *6*, 76. [CrossRef]

land

MDPI

Article

The Cultural Landscape Past of the Eastern Mediterranean: The Border Lord's Gardens and the Common Landscape Tradition of the Arabic and Byzantine Culture

Konstantinos Moraitis

School of Architecture, National Technical University of Athens (NTUA), 8A Hadjikosta Str., 11521 Athens, Greece; mor@arsisarc.gr; Tel.: +30-210-6434101

Received: 5 January 2018; Accepted: 21 February 2018; Published: 26 February 2018

Abstract: An evaluation of landscape tradition, in Near and Middle East area, could emphasize a profound past of agricultural experience, as well as of landscape and garden art. In reference to this common past, Byzantine and Arabic landscape and garden art paradigms appear to be geographically and culturally correlated, as proved by a Byzantine 12th century folksong, presenting the construction of a villa, with its surrounding gardens and landscape formations, in the territory of Euphrates River. This song refers to Vasilios Digenes Akritas or 'Border Lord', a legendary hero of mixed Byzantine-Greek and Arab blood; 'Digenes' meaning a person of dual genes, both of Byzantine and Arabic origin, and 'Akritas' an inhabitant of the borderline. At the end of the narration of the song, contemporary reader feels skeptical. Was modern landscape and garden art born in the European continent or was it transferred to Western world through an eastern originated lineage of Byzantine and Arabic provenance?

Keywords: Arabic landscape and garden art; Byzantine landscape and garden art; cultural sustainability; political sustainability; Twain-born Border Lord

---

1. Introductory References: The Western Interest for Landscape and Its Eastern Precedents

We ought to remark in advance that the present article is written by a professional design practitioner who believes, however, that space formative practices are not of mere technological importance. They may present, moreover, central cultural assumptions, in many cases correlated to the political identity of the societies in reference, as they not only depict but also support and enforce their political ethics.

It is under the conscious or the unconscious feeling of the political importance, which potentially inhere within landscape and garden art, that Western centralized 'civilization' and afterwards Western extended 'cultural' appreciation[1] recognized them as emblematic practices. They refer not only to practical interest for space formation, for the reclamation and cultivation of land, for agrarian economy; they also represent an overall volition for the rational control of nature, for its 'mastery and possession'[2].

---

1 We tend to use the term 'civilization' (or the adjective 'civilized') in order to describe societies with centralized organization and centralized production of knowledge, correlated to 'civic' societies principally. In comparison we use the term 'culture' (or the adjective 'cultural') in order to describe a non-necessarily centralized social identity, social behavior and social production in general.

2 As described by René Descartes: 'La science peut nous rendre comme maîtres et possesseurs de la nature – Science may turn us to be masters and possessors of nature' [1] (p. 84).

We could compare the previous Western need for environmental 'policy', with the expansive colonizing identity of modern Europe, with its political ethics. Thus we could also understand the need of Western historians to present the period of the first Western neoteric political formation, Italian Renaissance, as the natal period of landscape appreciation and landscape art. In contradiction to this approach, cultural geography clearly stated that no society could exist, not possessing landscape formative practices or, at least, a landscape perception for its place of living [2]. This generative affinity between social structure and landscape organizing activity, mental activity of cultural perception and interpretation or applied construction activity, may appear even more 'loudly', in the context of the Eastern Mediterranean and Near and Middle East[3]. We refer to a privileged geo-cultural territory, where the first development of agricultural prosperity coexisted with the first need for geometrical abstraction.

2. The Eastern Genealogy of the Western Landscape and Garden Art

We have just presented a central argument of our exposé, insisting on the fact that Western ethnocentrism tends, in many ways, to underestimate the impact of exterior contributions to the genealogy of neoteric Europe. However, it is clear that the input of organized knowledge from Arabic world presented important influences for Europeans, while the same seems to be true for the influences exerted by Byzantine culture as well. It is rather obvious for objective historic research that Byzantine scholars offered an important help to the first flourishing of the Italian Renaissance in Florence. Italian Renaissance could be regarded in this way as the immediate sequence of the Palaiologian Renaissance, a period of important achievements in culture, concerning the revival of ancient Greek wisdom in Byzantine Empire. We could therefore speak of Byzantine originated influences, while at the same time we should remind the imprint of the Arabic knowledge, introduced to Europe through the Iberian Peninsula. An even more detailed description could present a mixed Arabo-Byzantine body of influences as introduced to Europe, for example, through the famous Arabic libraries of Spain, where scriptures of Arab and Greek scholars co-existed.

Let us now return to the exact subject of our interest, the historic past of the landscape perception in the Eastern Mediterranean, in Near and Middle East. Let us now insist on the landscape description in poetry, in folksongs originated in the Middle Byzantine Ages and still in use in Greece and Cyprus. Let us now focus on the rocky landscape backgrounds of the Byzantine icon painting or on the still surviving garden examples of Alhambra. Then we could rather rush to conclude that the geographically extended and prosperous cultures of the past, that of the East Roman Empire, also known as Byzantium, and that of the Arabic kingdoms, could not be alien to landscape and garden art and, moreover, to an important landscape feeling, to an unconscious and in the same time conscious relation to landscape, which was afterwards devaluated by Western neoteric non-reference. We could even assert that a highly developed knowledge of landscape intervention existed in the Byzantine and Arab world, prior for sure to the analogous approaches of the Western Renaissance period; a knowledge that offered its pre-existent maturity to Westerners.

3. The Political Importance of the Re-Evaluation of the Neoteric Landscape History

What the present article is all about? Is it about the history of landscape, or is it about the importance of landscape formative techniques in general, of landscape aesthetics or environmental sustainability? We dare to admit that the principal aim of this article tends to surpass the previous issues. Its principal aim is to emphasize the need for cultural and political 'sustainability' of peoples in Eastern Mediterranean, in Near and Middle East; to insist on their need to evaluate their cultural

3 The landscape formative activities described in the Border Lord's folksong that will be presented principally refer to Euphrates region that means to the Middle East zone. However some of its references, as those concerning vegetal acclimatization procedures refer to a larger Eastern Mediterranean and Near East territory, to the extended territory of interconnection between Byzantine and Arab cultural groups.

and political identity correlated to their profound historic past. Even more, its principal aim has to do with the recognition of their age-long liaisons, of their cross-border bounds mutually correlating their knowledge as well as their tradition and cultural experience. It is under this ultimate political scope that indigenous cultures of Eastern Mediterranean, of Near and Middle East, have to reconsider their correlation to their landscape substratum; not on the 'shallow' prospect of touristic activity only, but on the deepest need for self-conscience and self-esteem.

Western political identity is intensively correlated, since the Renaissance, with landscape perception, with landscape and garden art. The upraising of the organized political control in Italian Florence is clearly expressed through the application of perspective geometry in the re-structure of urbanscape, or in 'extra muros' landscape formations. Baroque grandiose gardens were intimately linked to the promotion of the royal sovereignty in Louis XIV's France or in other European kingdoms. Moreover, the bourgeois innovatory regimes, the innovatory liberal political tendencies, as expressed in the British Isles in particular, were expressly correlated to the genesis of the English 'landscape architecture'. It was under those powerful political connotations that landscape perception and landscape art gradually developed their neoteric ideological allure; it was under those powerful political connotations that the genesis of the neoteric landscape perception and landscape art had to be principally and even exclusively European. In his *Dissertation on Oriental Gardening*, firstly edited in 1773, the English architect Sir William Chambers referred to the influences exerted to the English landscape architecture through Chinese garden examples [3]. The result was an immediate polemic against his commentary; the principal British space formative art that expressed the British political and social ethics, the British political liberal volition, could not be presented as 'imported' from abroad.

Three centuries after Chamber's proposal, ought we to re-write neoteric landscape history? Do we have to re-evaluate it for the profit of the 'peripheral' influences, which, anyway, used to be historically and politically central in the remote pre-Renaissance past?

4. A 'Border Lord' of Dual Origin and His Landscape Formative Activity

We have already commented the fact that Western ethnocentrism has in many ways underestimated the impact that Byzantine and Arabic influences exerted on the genealogy of modern Europe. It is probably because of this generalized devaluation that Byzantine landscape and garden culture has been seriously researched by Western scholars only recently[4]. Offering a comparative example we may comment that Marie Louise Gothein, a pioneer landscape historian, dedicated only six and a half pages to the subject of the Byzantine Gardens, in her famous book *A History of Garden Art,* while presenting the Islamic Gardens in 26 pages, focusing her interest on the Arabic examples [5] (Vol.1. pp. 137–168). In comparison Italian Renaissance and Baroque gardening was presented by her at the length of 484 pages [5] (Vol.1. pp. 205–459, Vol.2. pp. 1–230). However Gothein presented, both Byzantine and Arabic references in a common chapter, under the title 'Byzantine Gardens and the Countries of Islam', probably recognizing that Byzantines and Arabs shared a common 'ground' of cultural expression and, moreover, a common landscape experience concerning the borderline territories asserted by both populations.

The clearest paradigm of such a common Byzantine and Arabic landscape experience is probably offered by a Byzantine 12th century folksong, usually described by the name of its principal protagonist'Digenes Akritas' or 'Digenis Akritis' [6]. It refers to a fictional personality, a legendary Byzantine hero of mixed Byzantine-Cappadocian Greek and Arab blood, named 'Vasilios Digenes Akritas' or, in English translation, 'Twain-born Border Lord'. In a part of the song named by modern scholars with the indicative title "The Home, the Garden and the Tomb" [6] (pp. 63–66), the construction

4 In the fall of 1991 Dumbarton Oaks dedicated a roundtable to the subject 'Gardens and Garden Culture in Byzantium'. It was followed, in 1996, by a Colloquium under the title 'Byzantine Garden culture', which brought together a group of garden historians with scholars who were experts in Byzantine studies. The proceedings of the Colloquium were published by Dumbarton Oaks Research Library [4].

of a villa is presented, accompanied with its surrounding gardens and landscape formations in the territory of Euphrates River, with the addition of a bridge over Euphrates and a monumental tomb on the top of it. The reference to this part of the song, in correlation to the double ethnic origin of its protagonist, will be used as the central argument of our article, exemplifying the assertion that organized landscape formative activity, similar to what developed latter in 14th and 15th century in Europe, already existed in Eastern geo-cultural regions, in the territory of Eastern Mediterranean, in Near and Middle East. If historians agree that the folk song was already in extended use during 12th century then, the cultural experience described by it must be even earlier, possibly originating in the 11th century or even earlier in the 10th century. Extending our previous statement, we may remark in addition that this cultural experience appears to be a common production of cultural affinity between different ethnic groups inhabiting the territory, a production of dual origin at least, as Vasilios Digenes Aktitas himself was.

5. Presenting the Part of the Folksong Described as "The Home, the Garden and the Tomb"

'Vasilios Digenes Akritas', we have already mentioned it, are the Greek names of the 'Border Lord'. 'Vasilios' is the first name of the hero to whom the folksong refers, while 'Digenes' and 'Acritas' describe his identity, his personal origin and his social and military status. The adjective 'Digenes' means a person of dual origin and 'Akritas' an inhabitant of the borderline obliged, as it also happened in the previous historic example of the Roman armed 'limitanei'[5], to react as border guard, in the case of an enemy invasion.

The folksong in question has the form of an extensive epic narration, known under a general title identical, as we have also mentioned, to the name of the hero, 'Digenes Akritas'. Six manuscripts have been preserved dedicated to it, the oldest two held one in Escorial Library, in Spain, in a version of 1867 lines indicated with the letter E, and one in Grottaferrata Library, in Italy, in a version of 3749 lines indicated with the letter G. In the first part of the epic the lives of Digenes' parents are narrated; the way they met, and the way his father, an Arab Emir, was converted to Christianity after abducting and marrying Digenes' mother. Then the Emir resettled in 'Romania', in the lands of the Byzantine population of Greek origin, together with his people. The second part of the song discusses, often from a first-person point of view, Vasilios' acts of heroism on the Byzantine border. Finally, having defeated all his enemies Digenes builds, as described in the last part of the song to which we refer in particular, a luxurious villa by the river Euphrates, surrounded by an extended garden and there he peacefully ends his tormented life.

In a clear way this last part of the narration, entitled by the literary commentary with the description 'The Home, the Garden and the Tomb', may be regarded as evidence of an age-deep experience and organizational knowledge, concerning landscape and garden formation practices. It refers to an age-deep experience and knowledge established in Near and Middle East areas by people of mixed origin or by populations being in correlation to mixed cultural impacts. The above, seems to be the first important statement that our article has already emphasized. A second important statement refers to the maturity and complexity of the examined landscape formative example. The folksong presents an organized variety of practices, starting with a site evaluation previous to the beginning of the construction works. Thus it seems to offer a description of practical guidelines, similar to those proposed two centuries later in the *Re Agraria*, a guide for agricultural and landscape practices written in poetic form by Michelangelo Tanaglia in Renaissance Florence[6].

In Akritas' folksong a less detailed sequence of works is denoted, however it is described in an organized way. Following the evaluation of the site, land reclamation activities take place.

---

5 The Roman 'limitanei' as the Byzantine 'akrites', were settlers of the border Roman or Byzantine zones. The central government offered them the ownership of land and, in return, they ought to protect the empire from enemy invasion.

6 As Mariachiara Pozzana's describes, Michelangelo Tanaglia's Re Agraria, was written in poetic form, thus associated to Digeni's folksong presented in our article [7]. (pp. 120-137).

Afterwards the villa is built and the formation of the gardens follows. Transplantation of vegetal species from other remote countries is testified and also description of minor constructions is offered, as for example those concerning the decorative fountains of the garden. At the end of the song, the contemporary reader feels astonished. Was modern landscape and garden art born in the European continent during Renaissance period, or was it rather transferred to the Western world through an eastern originated lineage, related to the Byzantine and Arabic culture?

6. The Landscape Formative Activities in Detail; the Decision for the Construction of the Villa and the Practical and Aesthetic Evaluation of the Site

Let us follow now the presentation of an extended sequence of landscape formative activities, as presented in the previous folksong of Digenis', in the part of the song entitled 'The Home, the Garden and the Tomb' [6] (pp. 63–66). At the beginning of a song, we have a summary of the heroic feats of Digenis', and his decision to build his own house in the meadows.

'*... Then the whole Universe frightened by his power was.*
Many local chiefs under his control were,
And many chiefs of robbers were by Acritas killed, as well as many 'apelates'[7].
Thus no other concern for military attacks
Or for military defense had he anymore ... V.1615
And as no other concern for nice deeds had he,
He decided a house to have, a house to build in the meadows ... '

And the folksong continues by describing Digenis' effort to locate the proper site for the construction of his house, an effort correlated to the natural properties of the landscape, the irrigation possibilities and the rich flora of the place, but also with the aesthetic qualities, the 'beauty' of the water streams. The natural formation of the site may already be compared to the aesthetic elegance of an artificially 'composed' garden, of a 'paradise'; a term used to describe in Greek, during this period, not only the holy primordial place of the Scriptures, but also the artificial landscape formation, a garden. Thus we may answer to the arguments of the Western scholars, as for example to the Western centered opinion of Joachim Ritter, according to which the landscape perception has to do with the *Function of Aesthetics in the Modern Society*, of the Western society in principal [8] (pp. 33–97).

First of all, landscape feeling has not necessarily to do with conscious appreciation and description, or with the conscious 'distance', between the spectator and the objectified evaluated perception of his interest, created in the neoteric world. Otherwise we could not refer any more to a 'phenomenology' of landscape, in the Heidegger's use of the term, or to an immediate aesthetic appreciation of the un-cultivated mind in response to its perception. In any case, even in reference to the sophisticated Western mind of Ritter's exposition, the societies of the Western modernity were not the first to develop a 'res cogitans' examining and evaluating the landscape 'res extensa'. In Digenis' song, the hero visits a sequence of possible sites, where his house, a villa similar to those constructed three or four centuries later in Europe, may be erected accompanied by its surrounding garden. His examination has to do with practical needs and with an aesthetic evaluation as well, many years before the proverbial ascent of the Italian scholar and poet of Renaissance Italy, Francesco Petrarca, on Mont Ventoux [9]. Francesco Petrarca commonly anglicized as Petrarch, is often considered the founder of Humanism, and his landscape wondering from the top of the previous mountain, accompanied by a literary description, is also considered as the first 'civilized' [1] recognition of the landscape aesthetic qualities. In the Border Lord's song we may find descriptions of an analogous aesthetic admiration; though we

7 'Apelates': independent guerilla warriors of the Byzantine border zone.

refer not to a letter of a prominent scholar, as in the case of Petrarch's but to a folksong, in use by the vast majority of the inhabitants of average or even low education. We refer, however, to the inhabitants in a territory of extremely high cultural past. In comparison to the Western Middle Ages, they still live, at the time that the folksong appears, in a rather active cultural present that lasts until the period of the great destruction created by the European Crusades.

At the beginning of his landscape formative intervention the Border Lord inspects the possible sites of his future installation, finally arriving to the land of Euphrates (Figure 1).

'Thus to the nearby river-bank zone he went,
But no suitable place, for a person as noble as he, he found;
So to the land of Euphrates he decided to go. V.1620
... *In a place of meadows many trees he found,*
Standing around, thick shadow they offered,
While the beauty of water streams
Down from the mountain was coming. V.1625
Thus the site as beautiful as a paradise[8] *seemed'*

Figure 1. 'Thus to the nearby river-bank zone he went, but no suitable place, for a person as noble as he, he found; so to the land of Euphrates he decided to go ... In a place of meadows many trees he found, standing around, thick shadow they offered' (sketches by the author).

[8] The word 'paradise' is used as synonym to the word 'garden'.

7. Land Reclamation Activities and Landscape Architecture Interventions

Then the Border Lord continues with land reclamation activities, as a proof of an already developed experience of similar geotechnical interventions. It would be important to mention that big gardens and landscape architecture interventions, in European history—as in Vaux-le-Vicomte and Versailles Baroque gardens, in Dutch polder formation, or later in the English garden park of Blenheim—were also largely associated to the efficacy of analogous engineering projects; however, the description presented by the Border Lord's song is much earlier.

'So the river out of the meadow sent he,

And a garden of pleasure he actually formed . . .

. . . The four confluent-streams of the river he took,

For the irrigation of his whole dominion to use them' V.1635

The description continues with the presentation of the smaller parts of the water management infrastructure, the irrigating fountains (Figure 2, details n. 1 and n. 3) and the 'vivaria', a term describing in the song aquaria constructions, 'full of fishes'.

'Fountains, cast-out of iron he placed,

Enclosed parts to irrigate

(And) vivaria of wonder, full of fishes he made' V.1638

The landscape formative works continue with an activity rather common to the Western societies during the next centuries. We refer to the acclimatization of vegetal species introduced from distant places, as palm trees and the red St. John's wort, brought from Egypt and described in detail in the song.

'Then palms he planted in this garden V.1641

And out of Egypt the red St. John's wort they brought: V.1640

Its leaves green they are, but its flower red. V.1642

Long its root is, of fragrant wood,

And perfume full its fruits are,

Its branches red, spinning leaves around. V.1645

Snow-white juice out of the root springs out,

Rose-water smell, a sweet euphoria producing'

8. The Transition from Tradition Cultural Practices to Organized Knowledge: Landscape, Botany, Literary Expression

Let us quote a recent scientific article concerning the last verse of the song.

It indicates that 'studies have supported the efficacy of St. John's wort as a treatment for depression in humans' [10], an indication close to the descriptive words 'sweet euphoria', used in the song (Figure 2, detail n. 2). We must not be astonished for the medical knowledge of the East Mediterranean remote past historic societies. Byzantines and Arabs were excellent doctors and possessed deep knowledge of medical herbs. Thus we may find 13th century Arabic translations of *De Materia Medica*, a famous five-volume encyclopedia written much earlier under the ancient Greek title *Περί ύλης ιατρικής*—On Medical Material, by Pedanius Dioscorides, a Greek physician, pharmacologist and botanist of the first century A.D. This medical thesaurus was dedicated to herbal medicine and to relate therapeutic treatment, it was widely read for over 15 centuries, and seems to be highly respected by Byzantine and Arab physicians equally. A Byzantine manuscript of it with notes in Arabic[9], probably belonging to the library an Arab doctor of the 11th century, proves that knowledge of both languages, Greek and Arabic, was common at the territory we examine and that cultural and scientific exchanges were a continuous fact. It also proves the correlation to nature in a highly organized way, comprising technical capacity as well as scientific intentions, concerning for example developed landscape formative techniques and a fairly developed expertise in botany.

If we should like to extend our remarks about the refinement of the cultural practices, gradually developed form a state of the tradition and the folk-culture to the state of a centralized sophistication, corresponding to a highly elaborated expression, then we must also mention the changes in the composition of the song of Digenes' itself. While the text of Escorial manuscript appears to be closer to the features of the oral origins, the Grottaferrata manuscript is heavily marked by learned reworking. In both cases the manuscripts preserved, indicate the transformation of the immediate verbal 'tradition' to the written and afterwards to a more sophisticate re-evaluation. This cultural transformation from oral tradition to written, and then to elaborated expression, seems to be comparable to the transformation of the traditional agrarian practices to the developed land reclamation and to the landscape intervention, or to the transformation of the traditional use of herbal remedies to organized botanical, medicinal, and medical knowledge.

9. Art Imitating Nature and the Mechanical Replicas

If we return to the folksong we present, we shall find Digenes decorating his garden (or should use the word in plural; his gardens), with highly sophisticated devices usually described as 'automata'. Similar mechanisms, as we know, existed in the Abassid palaces in Bagdad as well as in the throne room[10] of the Byzantine emperor Theophilos, in Constantinople, in 10th century, as the Italian ambassador Liutprand of Cremona testifies[11] (Figure 2, detail n. 4).

'... All gold and all silver animals he erected,
Lions, leopards and eagles, partridges and fairies,
Springing out of their mouths and wings
Clear water, fragrant and crystal like. V.1655
All these on sublime fountains were placed'

It is normal to conclude that a culture, animating its gardens with mechanical replicas of living creatures springing out water out of their mouths, possessed a rather developed landscape art. In the 17th century, the famous French philosopher René Descartes proposed a Baroque garden fountain, in his book *Les Météores* [1] (p. 198). It seems to be, in comparison, a construction much simpler than the 'sublime fountains' of Digenes'.

Then the Border Lord continues with the winged inhabitants of the gardens and, having finished with other projects, he proceeds with the construction of a single vault bridge, over the river, with a monumental tomb on the top of it (Figure 2, n. 5). To describe the bridge, the narration uses again an adjective of aesthetic value; it is a 'handsome bridge' connecting the two banks of Euphrates and, at the same time, the earthly existence of the Lord with his future life in the heavenly garden of Paradise, the term being now used in our article with its religious connotation.

'Then golden cages he hanged on the tree branches
Nice parrots in them, chirruping and telling:

---

9 It exists in the library of the Georgian Monastery of Iviron or Iveron, an Eastern Orthodox monastery in the monastic state of Mount Athos in northern Greece.

10 For the use of wind powered automata in Bagdad see J. W. Meri's description [11].

11 For automata in Byzantium see G. Brett's description [12] (pp. 477, 487) as well as M. L. Gothein's references [6] (Vol. I, p. 141). In any case it seems that an ancient tradition of manufacturing automata is ascending from Greek antiquity and continues to be vivid in Jewish legends, as those concerning King Solomon's throne and in ancient Chinese references.

'Long live to the lord and to the lady of his desire'...
... And finally a handsome bridge he built, over Euphrates
A bridge of a single vault.
A court of four entrances was built over the bridge ...
... and inside a tomb there was,
For his own body to be buried'

Figure 2. **1** and **3**—*'Fountains ... enclosed parts to irrigate'*; **2**—*'and out of Egypt the red St. John's wort they brought'*; **4**— *'lions, leopards and eagles, partridges and fairies springing out of their mouths and wings clear water,'*; **5**—*'a court of four entrances was built over the bridge ... and inside a tomb there was'* (sketches by the author).

10. Conclusive 'Rhetorical' Questions and Conclusive Statements

For many Western scholars, the appreciation of landscape must be correlated with Renaissance, more precisely with Petrarch, who climbed on Mount Ventoux in order to enjoy the distant view, in a state of aesthetic delight [9]. In an analogous mode of reference, Western political interpretation of history correlates the emergence of the garden and landscape formative practices with the villas construction in the countryside surrounding Florence, in Toscana; with their landscape 'scenic integration' and their formal gardens.

If we accept this well-known historic approach, how can we then judge the legendary references to the gardens of the Border Lord? What have we to say about the paradigm of the Arabic garden, still preserving its lineage in Alcázar of Seville, in the Court of la Acequi in Generalife, in the Lions Fountain of Alhambra? Finally, how can we judge the need of the ethnic groups, in Eastern Mediterranean, in the Near and Middle East, to establish a contemporary political and cultural 'sustainability', in correlation to their history and to their ancestral landscape of origin?

Comments on the central ideas underlying the present paper, could probably criticize the previous conclusive questions and their tendency to put Eastern and Western evidences of landscape and garden art, into a 'competitive relationship'.

We ought to insist on the importance of this criticism, in order to explain that the previous tendency is not a negative obsession of the author. On the contrary, it seems to express a number of prominent Western intellectuals; Georg Simmel in his essay on the philosophy of landscape, *Philosophie der Landschaft* [13], Joachim Ritter in his essay on landscape and the function of aesthetics in neoteric society, *Landschaft—zur Funktion des Ästhetischen in der Modernen Gesellschaft* [8], we already mentioned him, or Ernst Hans Gombrich in his *Renaissance Theory of Art and the Rise of Landscape* [9]. All of them explicitly insist on the opinion that landscape appreciation is a matter of conscious response towards the qualities of the environment; of conscious response, which could be only produced in neoteric post-Renaissance societies. It is under the same ideological approach that the art historian Kenneth Mackenzie Clark referred to landscape painting, to *Landscape into Art* [14], insisting on the assumption that it firstly appeared in Renaissance depictions and attained its mature quality during the next centuries.

We could certainly excuse the previous approaches, through the explanation of their limited historical documentation or their restricted theoretical depth. Most likely they were not in touch with the 20th century doctrines of historical geography, of Carl Ortwin Sauer's 'cultural landscape' approaches for example [2]. According to them, no culture could be conceived outside place attributes, without cultural forms superimposed on natural landscape, without conscious or unconscious landscape references. Thus, conscious apperception is not 'sine qua non' causation for cultural landscape formations, while Pre-Renaissance place references and place intervention practices are not necessarily deprived of serious organizational landscape qualities. On the contrary, Arabic garden examples appear to be as formal as the Euclidean geometry gardens of renaissance and Baroque era, correlated to the appraisal of the Western conscious intellect. In an even more specific way, in the folksong just presented, a developed expertise of landscape reclamation techniques is revealed, described side by side to a an equally developed experience of botanic knowledge and flora acclimatization practices, to which we could safely acknowledge the quality of conscious interventions. We could equally reverse Clark's argumentation, by insisting on the arboreal depiction, as presented in the late Paleologean era frescoes in Pantanassa church of Mystras; or we could argue that landscape references on the background of the Byzantine icons are not 'primitive' or 'immature' representations in comparison to the 18th Dutch landscape painting examples. Under a more sophisticated statement of art hermeneutics, they could be accepted as representatives of an expressive 'schematization' much closer to the early 20th century abstract art, than to the 18th and 19th Western realistic tendencies.

However, the most revealing example of the 'competitive relationship' between Western neoteric landscape culture and its oriental precedents is probably the one concerning the public reaction against William Chamber's book, *A Dissertation on Oriental Gardening* [3], already previously mentioned.

Chamber's argumentation referred in his book, as we already stated, to the supposition of formative influences exerted to Western landscape art, especially to the 18th British landscape architecture, through examples of Asiatic origin, transferred to the European continent. Though the author's arguments referred to Chinese and not to Near or Middle East influences, the reaction to it was representative of the general desire of the Western societies, to appropriate landscape and garden art and promote them as a genuine Western cultural creation. Minor restricted elements of Arabic, Chinese, or Japanese provenance could be introduced, decorating European gardens with 'arabesque' stylistic elements, or with 'chinoiseries' and 'japaneseries' ornamental constructions; however, principal compositional trends had to be of Western invention. We shall not continue our commentary on this highly irritating subject. We shall only assert that Western landscape theory approaches, as well as Western landscape art hermeneutics, appeared to be cautious towards 'exterior' influences. Landscape art, in both forms of landscape architecture and landscape painting, appeared to be, for four centuries at least, representative of the high level of the Western political identity and everyday social ethics; it seemed thus inconceivable that their attitude, regarded as principally Western, could be produced through alien guidance.

Our primary concern in the essay presented was to prove that mature landscape apperception, as well as mature landscape formation techniques, already existing in the Eastern Mediterranean and Near and Middle East area, at least 400 years earlier than the Renaissance flourishing of the garden art in Tuscany and afterwards to Rome and the rest of Italy. We could of course try to locate even older important paradigms, going backward to Hellenistic and Roman gardens, to Macedonian and Pompeii frescoes. We shall insist however in Eastern Mediterranean and Near and Middle East 10th century, for three interrelated reasons.

The first one refers to the similarity of the Border Lord's folksong description, to an analogous poetic landscape formative guide, written during 14th century in Florence. We already mentioned it; it was Michelangelo Tanaglia's *De Re Agraria* [7]. However the Border Lord's garden is described in a folksong, and not in a poem written by an intellectual presenting himself as an agriculture and landscape formative techniques specialist. It refers therefore to an extended knowledge, shared not only by educated readers but also by a much larger cultural group of interest, in 10th century Eastern Mediterranean and Near and Middle East; by an extended cultural group that produced this landscape knowledge at least two centuries before the appearance of the folksong describing it.

The second reason refers to the fictional creator of the landscape works described in the song. The Border Lord is a fictional personality; it represents nevertheless the osmosis of two co-existing cultures, the Byzantine and the Arabic one, their shared common experience, their interchanged knowledge and practices. They could even claim, those two interconnecting cultural groups, a common origin, mixed genes, rather bringing them together than separating them. Landscape formation practices could be regarded as closely correlated to this cultural osmosis.

The third reason, justifying the general attitude of the present essay, could be correlated to a contemporary cultural and political statement. The Border Lord or Digenis Akritas, this fictional hero of double origin, may be described as one of the longest living personalities of the Hellenic heroic Pantheon. Until recently, folk songs commemorating his feats were common in the agricultural regions of Greece and Cyprus, and literary references of Greek modernity presented him as a personification of Hellenic ethnic continuity. However, the Border Lord was not of Byzantine origin only; on the contrary he was half Byzantine and half Arab. It could be an interesting occasion to look backward to Middle and Near East past and indicate to the contemporary inhabitants of this controversial geopolitical territory that, besides continuous antagonism, they also share a deep tradition of synergies, of synergies surpassing religious or ethnic differences, of synergies having probably to do with their common affection to their land and landscape substratum.

Conflicts of Interest: The author declares no conflict of interest.

References

1. Descartes, R. *Discours de la Méthode, suivi D'extraits de la Dioptrique, et des Météores*; Garnier Flammarion: Paris, France, 1966.
2. Sauer, C.O. The morphology of landscape. In *Foundation Papers in Landscape Ecology*; Wiens, J.A., Moss, M.R., Turner, M.G., Mladenoff, D.J., Eds.; Columbia. University Press: New York, NY, USA, 2007.
3. Chambers, W. *A Dissertation on Oriental Gardening*; Printer to the Royal Academy: London, UK, 1773; Available online: http://books.google.com (accessed on 20 December 2017).
4. Littlewood, A.; Maguire, H.; Wolshke-Bulmahn, J. *Byzantine Garden Culture*; Dumbarton Oaks Research Library and Collection: Washington, DC, USA, 2002.
5. Gothein, M.L. *Geschichte der Gartenkunst*; 1904. English Translation by Archer-Hind *A history of Garden Art*; Hacker: New York, NY, USA, 1979.
6. Aleksiou, S. *Vasilios Digenis Acritis and Armouris' Song*, Greek ed.; Ermis publ.: Athens, Greece, 1985.
7. Pozzana, M. Agricoltura e orticoltura nella Toscana del Quattrocento. In *Giardini Medicei. Giardini di palazzo e di villa nella Firenze del Quattrocento*; Luchinat, C.A., Ed.; Federico Motta Editore: Milano, Italy, 2000.
8. Ritter, J. Landscape and the aesthetic appreciation in Modern society. In *To Τοπιο—The Landscape*, Greek ed.; Simmel, G., Ritter, J., Gombrich, E.H., Eds.; Potamos publ.: Athens, Greece, 2004; pp. 33–97, First published as *Landschaft—zur Funktion des Ästhetischen in der Modernen Gesellschaft*. Aschendorff Verlag: Münster, 1963.
9. Gombrich, E.H. Renaissance Theory of Art and the rise of Landscape. In *Norm and Form. Studies in the Art of the Renaissance I*; Phaidon Press: London, UK, 1966; pp. 107–121.
10. Klemow, K.M.; Bartlow, A.; Crawford, J.; Kocher, N.; Shah, J.; Ritsick, M. Chapter 11: Medical Attributes of St. Joh's Wort (*Hypericum perforatum*). In *Herbal Medicine Biomolecular and Clinical Aspects*, 2nd ed.; Benzie, I.F.F., Sissi, W.G., Eds.; CRC Press: New York, NY, USA, 2011.
11. Meri, J.W. *Medieval Islamic Civilization: An Encyclopedia*; Routledge publ.: London, UK, 2005.
12. Brett, G. The Automata in the Byzantine 'Throne of Solomon'. *Speculum* **1954**, *29*, 477–487. [CrossRef]
13. Simmel, G. Philosophy of landscape. In *To Τοπιο—The Landscape*, Greek ed.; Simmel, G., Ritter, J., Gombrich, E.H., Eds.; Potamos publ.: Athens, Greece, 2004; pp. 9–31, First published as 'Philosophie der Landschaft,, in *Die Guldenkammer II*, Bremen 1913.
14. Clark, M.K. *Landscape into Art*; Icon Editions: New York, NY, USA, 1979.

land

Article

Landscape Archaeology and Sacred Space in the Eastern Mediterranean: A Glimpse from Cyprus

Giorgos Papantoniou [1,*] and Athanasios K. Vionis [2]

[1] Research Training Group 1878: Archaeology of Pre-Modern Economies, Abteilung für Klassische Archäologie, Institut für Archäologie und Kulturanthropologie, Rheinische Friedrich-Wilhelms-Universität Bonn, Lennéstr. 1, D-53113 Bonn, Germany

[2] Department of History and Archaeology, University of Cyprus, P.O. Box 20537, 1678 Nicosia, Cyprus; vionis@ucy.ac.cy

* Correspondence: papantog@uni-bonn.de

Academic Editors: Ioannis N. Vogiatzakis, Theano S. Terkenli, Maria Gabriella Trovato and Nizar Abu-Jaber
Received: 6 May 2017; Accepted: 9 June 2017; Published: 14 June 2017

Abstract: This article aims to raise issues for discussion about the change in the use and concept of sacred landscapes, which were originally constructed in the era of the Cypriot kings (the *basileis*), but then continued to function in a new imperial environment, that of the rule of the Ptolemaic *strategos* and later of the Roman *proconsul* and the various Christian bishops. Our archaeological survey project in the Xeros river valley, titled 'Settled and Sacred Landscapes of Cyprus', reveals that these new politico-economic structures were also supported by the construction of symbolically charged sacred landscapes. Thus, while outlining the long history of the island as manifested from the diachronic study of Cypriot sacred landscapes, we identify three pivotal phases: first, the consolidation of the Cypriot polities and the establishment of a 'full' sacred landscape; second, the transition from segmented to unitary administration under the Ptolemaic and Roman imperial rule and the consolidation of a more 'unified sacred landscape'; and finally, the establishment of a number of Christian bishoprics on the island and the movement back to a 'full' sacred landscape. Moving beyond the discipline of Cypriot archaeology, this contribution aims to serve as a paradigm for the implications that the employment of the 'sacred landscapes' concept may have when addressing issues of socio-political and socio-economic transformations. While it is very difficult to define or capture the concept of landscape in a pre-modern world, it offers a useful means by which to assess changing local conditions. We have also attempted to situate the term in archaeological thought, in order to allow the concept to become a more powerful investigative tool for approaching the past.

Keywords: Cyprus; landscape archaeology; sacred space; political power; economy; religion; ideology; ancient sanctuaries; churches

---

1. Introduction

Landscape archaeology is currently hugely popular. However, there are problems with defining the term 'landscape'. This is partially because the term is used in various ways, not only between different disciplines but also within each discipline, which forces us briefly to analyze the concept in an archaeological context. It should be acknowledged that the term has also been used very freely in archaeological writings. Many scholars have used the term as a 'fashionable gloss' for survey studies or an improved 'site-catchment analysis' [1,2]. One should, however, try to understand the term in its broader sense.

The concept of landscapes encompasses temporality, spatiality and materiality; therefore, much can be said about human responses to the changing conditions of life over time through its study. The term is concerned both with the conscious and the unconscious shaping of the land and the processes of organizing space, and involves interaction between the physical environment and human presence. As Matthew Johnson notes, "the two senses of the term 'land-scape' are important here: not just the land, but how it is viewed or mentally constructed" [3].

Contemporary approaches to landscape archaeology include a broad range of archaeological issues, such as social organization, rural economy and sacred space, trying to extract agency, ideology, meaning, memory, identity, social order, morality and social transformation from landscapes. While the last decades of landscape archaeological research in the Eastern Mediterranean (and in the Mediterranean region in general) have seen an intensification of interest in bringing together geographic (concrete and descriptive forms concerned with determining the nature of and classifying places, as well as establishing the links between them) and sociological 'imaginations' (an aspiration to explain human behavior and activities in terms of social process abstractly constructed) [4], some archaeologists have seen landscape as "the arena in which and through which memory, identity, social order and transformation are constructed, played out, re-invented, and changed" [5]. Landscape then, can be seen as the 'arena' for social agency [6].

Cypriot sacred landscapes are rarely examined in relation to political power, political economy and ideology. Such landscape perspectives are absent from studies of the crucial transitions from the Cypro–Archaic and Cypro–Classical city-kingdoms to the Hellenistic period, from the Hellenistic to the Roman era, and from Roman times to early Christianity. Acknowledging the potential of landscape studies to provide a major source of new interpretations on the *longue durée* (the long-term approach to the study of the past, employed by the French *Annales* School of history) this contribution should be regarded as a dynamic attempt to re-work and re-experience not only the Cypriot sacred landscapes through various transitional phases, but also to further illuminate the political and socio-cultural histories of the Cypriot city-kingdoms, the Hellenistic, the Roman and the Late Antique periods when viewed independently (Figure 1). We demonstrate how a diachronic approach to Cypriot sacred landscapes, which includes ancient sanctuaries and early Christian basilicas can open new interpretative windows, impossible to reach without such a comparative approach. Late Antiquity, in particular, is a crucial period in the development of Cypriot sacred landscapes as again, similarly to the beginning of the Archaic period and the consolidation of the Iron Age polities, we move from a 'half-empty' to a 'half-full' and later to a 'full' sacred landscape.

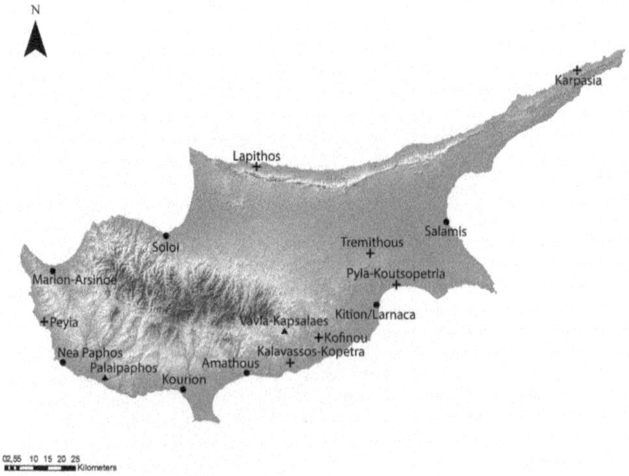

Figure 1. Map of Cyprus with places and archaeological sites mentioned in the text. Circle: ancient city; triangle: ancient cult-site; cross: Christian site (digital geological data courtesy of the Geological Survey Department, Republic of Cyprus; map drafted by A.K. Vionis).

2. Trends in Landscape Archaeology and Sacred Landscapes

Landscape archaeology has followed the main trends of the theoretical developments of archaeology by moving through a number of stages, though they have not been sequential. Recognizing the risk of oversimplification, Tony Wilkinson defined three broad strands of landscape archaeology. Firstly, the 'cultural-historical' approach (or the school of landscape history) draws on historical documents, archaeology and the landscape itself; secondly, 'processual' approaches embody a more 'scientific' methodology and include archaeological surveys, off-site and quantitative studies, catchment analysis, settlement archaeology and various ecosystem approaches; finally, 'post-processual' approaches are a reaction to the processual approach, and include phenomenological, ideational and symbolic/religious landscapes [7].

Scientific or functional approaches, under the strand of 'processual' archaeology, have usually predominated, inferring social and economic dimensions of a range of spatial frames and statistical models [8]. However, it was felt that the land remained a neutral and passive object, used by people, but otherwise relatively detached from them [8]. Considering all the different approaches to the study of past landscapes, it should constantly be kept in mind that places and landscapes have been formed by the very act of living. The human factor and involvement, therefore, are key concepts that should not be underestimated or overlooked. Moreover, human activity and landscapes' structure and temporality are vital issues, directly associated with the concept and perception of landscape [9]. Already by the 1970s, 'post-processual' and 'post-positivistic' philosophies, humanistic concerns and calls for social relevance, built from existentialism, structuration, Marxist thought, feminism, idealism, phenomenology, and interactionism, were recast "as matched participants in [a] perpetual dialectic of mutual constitution" [8]. Today the most prominent notions of landscape archaeology emphasize its socio-symbolic dimensions: "landscape is an entity that exists by virtue of its being perceived, experienced, and contextualized by people" [5]. Landscape is never inert; people are directly associated with it, re-work it, appropriate it and contest it [10]. In addition, the theory has moved on considerably to include 'ecological' and 'co-production' approaches in a more holistic way [11,12].

Landscape archaeology, therefore, has the potential to be truly unifying, bridging the gap between scientific or positivistic archaeologies and those that approach it from the perspective of social theory or the humanities [13]. There is undoubtedly a need for an integrated approach in which all the approaches mentioned above are taken into account. Such a holistic approach and interpretation, which regards landscape as a reflection of society and as an expression of a system of cultural meaning, and which seeks to read the materialization of ideologies on land and monuments, is currently applied to the reading of Cypriot sacred landscapes as part of the Research Network's 'Unlocking Sacred Landscapes' (UnSaLa) project and the archaeological project 'Settled and Sacred Landscapes of Cyprus' (SeSaLaC).

The term 'sacred landscapes' has been chosen in acknowledgement of the inspiration provided by Susan Alcock's work; by using this term in her examination of the Hellenistic and Roman sacred landscapes of the Greek world, Alcock shows that the relationship between religion, politics, identity and memory was more intimate and more involved than has often been assumed [14–18]. She regards sacred landscapes emerging:

> "... as both culturally constructed and historically sensitive, immensely variable through time and space. Far from being immune to developments in other aspects of human life, they can reflect a very wide cultural and political milieu. Yet they also provide more than a simple mirror of change by their active participation in the conditions of social reproduction". [14]

The investigation of 'ideational' or 'associative' landscapes, where people associate features in the natural and built landscapes with their own memories, meanings or emotions [5,16], is particularly relevant to sacred landscapes [6]. 'Ideational', as Bernard Knapp and Wendy Ashmore argue, is far less linked to an articulated system than the terms 'ideology' or 'ideological'; therefore, it can also be used to embrace sacred as well as other kinds of meanings attached to and embodied in landscapes [5].

The concept of memory is crucial in the process of socializing landscape and naturalizing cultural features in the land. It is created by the repeated movement of the body throughout the landscape. Barbara Bender regards landscape as a process that is "intensely political, a way of perceiving, experiencing, and remembering the world" [10]; landscapes not only shape but are shaped by human experience [8]. Christopher Tilley's influential study is concerned explicitly with phenomenology of landscape as an experience [19]. The experience is 'synesthetic', "both creating and engaging a narrative linking the body—individual and social group—with the land" [8]. The movement of the body through space is crucial, and as it provides people with a particular way of viewing the world, it has important implications for the maintenance of power relations [19,20]. By controlling the way people move through space, it is possible to reproduce dominant perspectives on the world [19]. Robert Johnston sees landscape as existing through two different understandings of 'perception': in the first, perception acts as a filter on the real world; in the second, it is a process through which people understand the world [21]. In studying landscapes, perception cannot be ignored and it should be acknowledged that perception is not beyond archaeological analysis [2].

Questions about ascribing meaning to landscapes and issues of social mechanisms by which meaning is attached, as well as the range of meanings that can be encompassed should be raised [8]. Meaning is usually attached through memory and ritual. However, memories and meanings are created afresh from generation to generation and differ between individuals. As Ashmore notes, "prominent among the meanings of landscape are power and identity, variously defined and expressed in sundry forms" [8]. As landscape delineates memory and declares identity, the land itself plays a fundamental role in the social and cultural order and in human relations. Further, "as a community merges with its habitus through the actions and activities of its members, the landscape may become a key reference point for expressions of individual as well as group identity" [5]. The transformation of landscapes has been associated with the transformation of the social order, coming from short-term events (socio-political time) or medium-term cycles (socio-economic time). As Knapp and Ashmore note, since landscapes embody multiple times as well as multiple places, they consequently materialize not only continuity but also change and transformation [5]. Landscapes are perpetually under construction, which is why an enduring theme in recent archaeological thought has been the reading of social power from those modified landscapes [22].

John Cherry has emphasized the need to bring into a closer dialogue the various approaches of landscape archaeology [23]. Survey reports should be combined with excavation reports, political history and notions of recent 'archaeologies of landscape' [5]. Emphasis should be given to "the process of re-interpretation and re-working of dynamic landscapes whose changing appearance communicates cultural values and is charged with meaning" [23]. The study of Cypriot sacred landscapes, therefore, may become a significant interlocutor, which stimulates the understanding of the broader political, economic, and cultural space.

3. Cypriot Sacred Landscapes of Ancient Cyprus (Archaic-Roman Times)

The study of Cypriot sacred landscapes within the *longue durée*, their transformations and their possible change of meanings, reinforce current interpretations suggesting that the extra-urban sanctuaries played an important role in the political setting of the city-kingdoms, which transformed over time. Moving from the period of many independent *basileis* to the island-wide rule of the *strategos*, it has been discussed how Hellenistic 'urbanization', settlement patterns, social memory and politico-religious ideology present a picture of a more unified socio-political sacred space. As shown in Figure 2, by the Roman period many Cypriot extra-urban sanctuaries were deserted and only a few of them remodeled and enlarged. If we are going to understand this phenomenon, we have to move well beyond the Roman period (before and after) in order to insert the development of Cypriot sacred landscapes within their individual and contextual insular development.

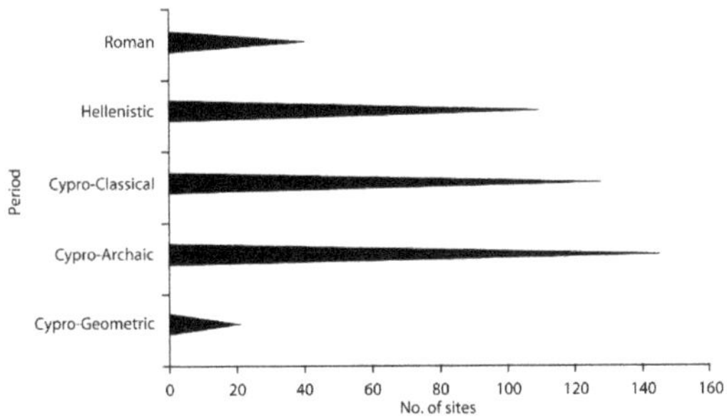

Figure 2. Density of sanctuary sites from the Cypro-Geometric to the Roman period (graph and data by G. Papantoniou).

Old excavation of extra-urban shrines of the Archaic and Classical periods has produced evidence that has also been confirmed by recent systematic excavation activity [24], and which highlights the role of the Cypriot Iron Age sanctuary as a focus of wealth disposal and economic control in the community [25,26]. There was evidence of the segmentation of space, consumption of food and drink, industrial activities, large-scale storage and display, and the disposal of votive images related to royal ideology. While urban sanctuaries become religious communal centers, where social, cultural and political identities are affirmed, an indication of the probable use of extra-urban sacred space in the political setting of the various city-kingdoms has usually been observed [27–42]. One of the present authors has argued in print on several occasions that the distribution of these sanctuaries across the landscape served as a map for a socio-political system, which provided a mechanism for the centralized Archaic and Classical city-kingdom authorities to organize and control their peripheries. Environmental and Geographic Information System (GIS) analyses that were employed for the first time in the history of research on Cypriot sacred landscapes by UnSaLa and SeSaLaC, reinforced this argument about the territoriality of Iron Age Cypriot sanctuaries. Next to other archaeological evidence and GIS analyses that support the relation of the Vavla-Kapsales sanctuary with the polity of Amathous, for example, the results of Cost Surface Analysis, which calculates the walking distance between places, also strengthened this interpretation (Figure 3) [41].

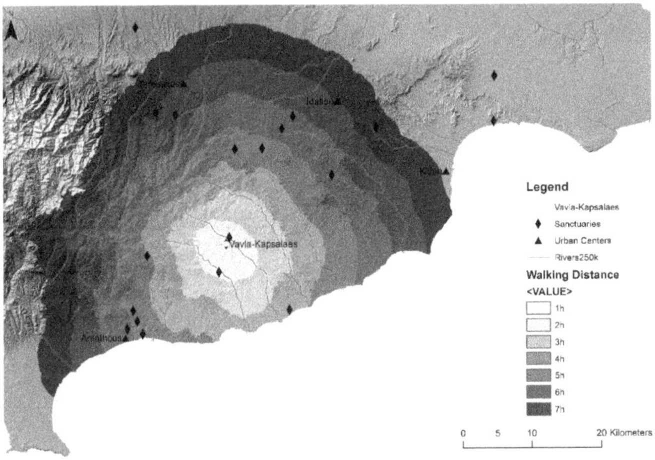

Figure 3. Cost Surface Analysis results from the Vavla-Kapsales ancient sanctuary (digital geological data courtesy of the Geological Survey Department, Republic of Cyprus; GIS analysis by N. Kyriakou).

The transformation of Hellenistic political topographies, and the passing of Cyprus from segmented to unitary, colonial administration under a foreign general (the *strategos*) brought a marked urban and extra-urban change [37,40]. In a unified state that offered unlimited access to inland resources, official emphasis was placed primarily on urbanized and strongly Hellenized coastal centers for political, military and economic reasons. The coastal cities, such as Nea Paphos, Marion-Arsinoe, Kourion, Amathous, Soloi and Salamis (Figure 1), undoubtedly mirror Ptolemaic strategic interests in coastal port bases and settlements as well as maritime activity and power [43]. Although Cyprus was ready to adapt Hellenic forms of administration, it was indeed Ptolemaic rule (and particularly that of the second and first centuries BC) that established city life and institutions, such as *boule*, gymnasia and theatres, in accordance with the cities of the Hellenistic world elsewhere [44,45]. While we can usually assume the shape of the Hellenistic cities through the epigraphic record, the current remains of all the coastal urban centers date to the Roman period, and they fully adhere to the monumental character of a standard Roman city.

The gravitation of people towards these coastal cities was of greater historical significance. Several archaeological surface surveys on the island have noted a full Hellenistic and Roman countryside settlement pattern, which was followed by a general contraction from the second through the fourth centuries BC [46]. According to Marcus Rautman [46], rural settlement began to decline around the second century AD, as demonstrated by published survey data in Cyprus and elsewhere; evidence of Severan prosperity, which often is taken to represent the apogee of Roman Cyprus, is overwhelmingly urban and may have come about at the expense of the countryside. However, the settlement patterns of the transitional phases are not entirely consistent, and new regional archaeological survey projects need to address those issues, exploring pastoral, agricultural and other economic activities and how these are related to the various political situations and to the siting of sanctuaries. We would preliminarily observe however, that the widespread abandonment of the extra-urban shrines, no less than the elaboration of public cults in the cities, set the stage for profound social and religious reassessments [46] that go back to the Hellenistic period, and have to be studied within the context of the transition from segmented to unitary government and administration.

Moving to continuity and abandonment in sacred landscapes, the importance of memory is a crucial factor [14,18]. It is important to see sacred landscapes not simply as constructs but also as a complex and dynamic reaction to Ptolemaic, and later to Roman incorporation, and to investigate what was remembered, when, where and how. The continuity or discontinuity of the extra-urban cult activity should be keyed to multi-polar power relations and memory trends. Local and non-local elites, in their effort to define or redefine their relationships with the land for political and economic reasons, or in order to naturalize or legitimate authority, could have used sanctuaries to demonstrate their status. As the epigraphic evidence reveals, new social divisions and affiliations within Cypriot society as a whole, but also within individual communities—intertwined with other sources of power—were also promoted through the very agency of cult. However, during the Hellenistic and Roman periods, the need for political elites to define the link between territory and city, for other than administrative or economic reasons, should have become less and less important. If extra-urban sanctuaries played a frontier or liminal role in the perpetuation of city-kingdom identities, now, under a new unified political organization, they eventually lost their territorial significance.

The dedication of monumental statues in many extra-urban sanctuaries by the Cypriot elite, imitating Ptolemaic prototypes—but at the same time adhering to the long typological and ritual Cypriot traditions—also reveals continuity in cult activity and traditions. The insertion of Hellenized and portrait-like features into statues whose general format remained strongly Cypriot might suggest a controversial ideological move [37,47,48]: not only the incorporation of Ptolemaic ideas (and ideals) into the Cypriot mentality, but also the accommodation of Ptolemaic rule into a Cypriot context, i.e., the incorporation of Cypriot ideas (and ideals) into the Ptolemaic ideology. Within the material record, resistance could be expressed in a covert manner, and involve the continuation of religious practices or the maintenance of a traditional material culture. Terms such as 'resistance accommodation' and 'resistance adaptation' have been widely used in contemporary archaeological literature to indicate that resistance was not an either/or proposition, but rather an ongoing, subtle and usually muted process [49]. In time, and when reaching the Roman period, many of the elites of the city-kingdoms would have disappeared or been suppressed. By the first century AD, therefore, when a more unified and centralized politico-religious system seems to have been well established, the tradition of dedicating limestone portrait-like sculptures in extra-urban sanctuaries eventually died out, along with the majority of the long-standing extra-urban sanctuaries. After the early Roman period, we find hardly any limestone 'portraits' in a Cypriot sanctuary; they all belong to the funerary sphere, carved on a number of grave reliefs [50–52]. The material culture of the Cypriot sanctuaries of the Roman period clearly suggests that identities and modes of self-expression had significantly shifted and transformed.

Religion represented a close linkage between local cult and local identity. Although we cannot simply speak of many cultural identities in Iron Age Cyprus, the shifting from many political city-identities to one had consequences, such as the interruption of promoting particular local cults. In addition, just as people moved in the landscapes, creating, modifying, destroying or abandoning places or institutions, in the same way their identity is defined, re-defined or restricted. Such an embodiment could enable us to define how people transformed their identities through landscapes, and at the same time how landscapes themselves were transformed, adapting to the new socio-economic relations, as well as to the new socio-political identities and memory realities. Changes that occurred on different levels in Hellenistic and Roman societies influenced the religious sphere, cult practice and consequently the sacred landscapes. On the other hand, sacred landscapes themselves should not be seen as a passive reflector of social practice, but rather as an active expresser of it; functioning under a new unified politico-economic system, sacred landscapes eased the transformation of human identities and perception of space.

The image of the 'Cypriot Goddess' is the paradigmatic example for illustrating the transformations that occurred as the island moved from segmented to unitary government and administration, and the complexity of that process. Her artistic representations on sculpture and terracotta figurines, for example, show that by the Hellenistic, and later the Roman period, she was fully Hellenized, conforming to the iconography of Greek Aphrodite [40,53,54] (Figure 4). Moving from stylistic to cultic analysis, however, is simply impossible without considering the fact that some local particularities in her cult survived well beyond the end of the Hellenistic period. The most illustrative example comes from the 'archaic' cult place of Aphrodite in Palaepaphos, which under Ptolemaic and Roman rule was developed into a pan-Cyprian sanctuary and where strong epigraphic evidence for the practice of the ruler and imperial cult exists [55]: not only the architecture of the sanctuary remains close to the traditional Cypriot *temenos* from the Late Bronze Age to the end of the Roman period, but also the cult statue of the goddess keeps the aniconic shape of a *baetyl*.

Such continuities should be viewed in relation to both the local cultural identity and the character of politico-religious agency and ideology, which through various accommodations and transformations, seems to have reproduced the established socio-cultural norms. Nonetheless, as well-documented evidence from epigraphy or the diachronic study of cult in excavated sanctuaries—such as those of Palaepaphos and Amathous—reveal, a more unifying reorganization of cult can be noted during the Hellenistic and Roman periods [39]. Changing economic conditions within the Cypriot cities under a unitary government and administration also entailed some significant changes in financing, and as a result, in the sociological structure of their religion. In the Hellenistic, and especially in the Roman periods, financial management eventually shifted from the city-state to a more unified and centralized control. The most important bearer of a unifying ideology should have been played by the *Koinon Kyprion* (Union of Cypriots), dedicated to the promotion of the Ptolemaic, and later the imperial cult [44,55,56].

Figure 4. Terracotta figurine of semi-naked Aphrodite resting on a pillar from Amathous, AM 949, Limassol District Museum. Image© The Department of Antiquities, Cyprus (courtesy of the director).

Among the large distribution of settlements in the various published archaeological survey projects and in the Cyprus Survey Inventory, one of us has located 23 sites that might have functioned as a sanctuary *ex novo* in the Hellenistic period [37]. Next to the Ptolemaic official attention towards old traditional urban sanctuaries [55,57–59], Greek-style temple architecture is added to the Cypriot sacred landscapes quite late in the Hellenistic period, not in the extra-urban environment of Cyprus, but in the direct environs of the major urban centers, where Ptolemaic power and cult would have been practiced more markedly. On the other hand, during the early Hellenistic period, sanctuaries *ex novo*, such as that of Soloi-Cholades (Figure 5), with strong allusions to Ptolemaic cult, were built following the traditional Cypriot architecture of the Iron Age Cypriot *temenos*, probably drawing on the existing religious sentiment and cult [60].

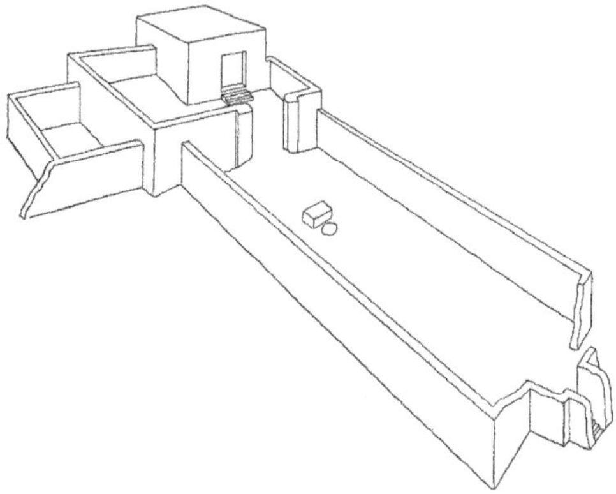

Figure 5. Soloi-Cholades sanctuary, reconstruction sketch of Hellenistic temple A [37].

As mentioned above, by the Roman period many inland Iron Age extra-urban sanctuaries were deserted and only a few coastal sanctuaries remodeled and enlarged (Figure 2); some of those sanctuaries, such as that of Apollo at Kourion and Aphrodite at Amathous, also received monumental podium temples. While excavation and survey projects confirm that an *ex novo* foundation of sanctuaries was rare in the Roman period [61], the use of pre-existing extra-urban sanctuary sites was visibly reduced. Only 40 possible sanctuary sites (including urban and extra-urban sites) preserved evidence of cults in the Roman period, and these sites included important 'time-honored' sanctuaries in the environs of urban coastal centers, such as the sanctuaries of Apollo *Hylates* at Kourion, of Aphrodite at Palaepaphos and Amathous, and of Zeus at Salamis. It seems that the Romans invested in rebuilding and temple constructions, usually at those same primarily urban sites as their Ptolemaic predecessors.

Summarizing the evidence presented above, it becomes clear that state revenues, financing for state festivals, and the building and upkeep of sanctuaries went towards more prestigious, high-status urban sanctuaries, such as those of Aphrodite in Palaepaphos and Amathous, or of Zeus in Salamis; this helped created a more unified national politico-religious identity. The primacy of these sanctuaries by the Roman period was confirmed by Tacitus (*Annales* 3.62), who reminded us that the Senate confirmed their rights of amnesty in 22 AD. Over the next 200 years, these cults increasingly became associated with the island's identity as a Roman territory. Images of the Palaepaphos sanctuary and the standing figure of Zeus Olympios appear both singly and paired on coins issued by the *Koinon Kyprion* for the first through the third century AD. The heavy promotion of these primary shrines, together with temples of the imperial cult at Nea Paphos and Kourion, for example, may be seen as part of Roman efforts to unify and consolidate the island province [46]. Following the Severan period, we know very little regarding Cypriot sacred space during the third and fourth centuries AD. The stratigraphy, material culture, and architecture related to the post-Severan sacred landscapes remains to be identified, published and sufficiently analyzed. We hope that scholarship will soon manage to fill or explain the gap we currently face during the third and fourth centuries AD.

4. From Roman Times to Christian Late Antiquity

Following the Severan period and the social transformations taking place during Late Antiquity (late fourth to middle seventh centuries AD), the urban temples of Cyprus eventually start declining. A weakened economy, and the subsequent imperial neglect during the third century AD, contributed to fundamental social realignments and dramatic ideological shifts [46]. Thus, town councils and magistrates did not maintain declining sanctuaries, which eventually closed because of earthquake damages during the mid-fourth century AD. The official establishment of Christianity, the economic prosperity that Cyprus started enjoying, and the shift of political control (at the local level) from imperial families to Christian elites and bishops contributed to the transformation of the sacred townscapes and landscapes of Cyprus [62–64]. Cypriot bishops worked in the shadow of damaged temples at the urban environments of Paphos, Amathous, Kourion and Salamis, resacralizing space by building cathedrals and other large basilicas in renovated parts of their flourishing cities [46]. During Late Antiquity, settlements and new cult sites dating to the fifth century AD, such as the basilicas at Karpasia, Lapithos, and Tremithous (Figure 1), document both the expansion of rural settlement and the successful Christianization of the countryside [46].

SeSaLaC is currently testing the above framework through a recently initiated archaeological surface survey project in the area of the Xeros river valley in the Larnaca district, with the modern village of Kofinou lying in its center (Figure 6). Preliminary GIS mapping and analyses of Cypriot Late Antique churches in relation to the road networks and arable land [65] aim to reveal—similarly to the function of ancient sanctuaries—the function of these countryside basilicas in the context of economic and symbolic landscapes. The analyses confirm that Christian basilicas are found in association with rural establishments (e.g., villages), local central places (e.g., towns, agro-towns and ports or coastal *emporia*), and other significant economic and communication nodes (e.g., road networks, rivers and agriculturally rich areas). The basilica churches at Kalavassos-Kopetra for instance (Figure 1), constructed in the sixth and seventh centuries AD on a hill in the middle of the Vasilikos valley, acted as regional economic nodes at a central point for the collection and distribution of imported and local products [65,66].

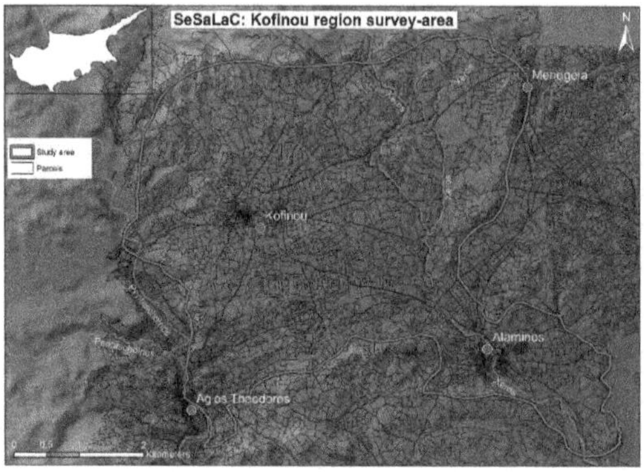

Figure 6. The survey area of the 'Settled and Sacred Landscapes of Cyprus' Project (SeSaLaC) in the Xeros valley, Larnaca District. Image© SeSaLaC (digital geological data courtesy of the Geological Survey Department, Republic of Cyprus; map drafted by V. Trigkas).

Our investigation in the region of Kofinou so far has indeed provided evidence for a much more intensive Late Antique settlement activity (in comparison to the preceding Roman era) around an Early Christian basilica (dedicated to *Panagia*, Virgin Mary) dated to the late sixth century AD (Figure 7). The archaeological evidence, topographic parameters, extensive surrounding agricultural territory and comparative evidence from other excavated and surveyed archaeological sites suggest that in Late Antiquity the site of Kofinou, around the church of Panagia, played a central role within its 'settlement chamber' or micro-region, coincidentally overlapping with our survey area [67,68]. Looking at population figures based on excavation and survey evidence, the site of Kofinou itself should have provided housing to approximately 250–300 families during its maximum size in the sixth and seventh centuries AD, when the built space around the basilica, according to surface ceramic scatters, reached almost 13 ha. The excavated basilica must have functioned as the focal point of the settlement, standing at its approximate center, dominating its immediate environs and marking a primary approach to the site.

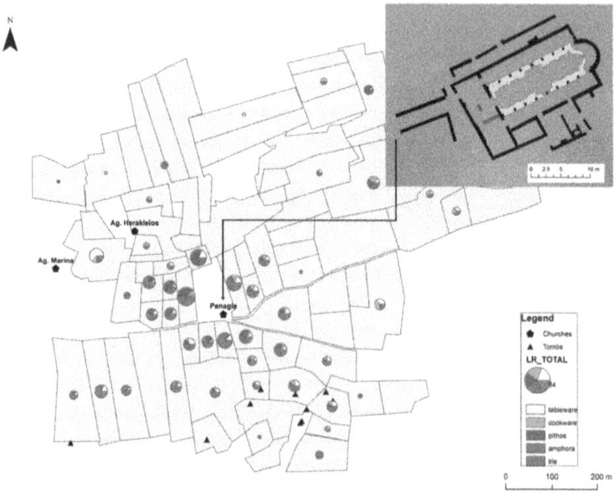

Figure 7. Late Antique surface ceramic concentration (by pottery types) around the Early Christian basilica of *Panagia* in the region of Kofinou, with the architectural plan of the church. Image© SeSaLaC.

The extent of the site, the rich ceramic evidence dated to Late Antiquity, and the presence of an important monument of Christian worship in an otherwise extensive and mostly fertile agricultural region can only point to the status of Kofinou as a second-rank settlement, and as the main habitation site of the micro-region of the Xeros valley. Such secondary settlements in the countryside had a major role to play as local centers, that is, as important loci within the territory of their 'settlement chamber', acquiring an important role in agricultural production, processing and distribution of goods, and sometimes administrative functions as well. If, then, Kofinou comprised a second-rank settlement, which we would define here as an 'agro-town', one needs to identify the closest primary center, or regional central place, and other minor rural establishments. Although this remains guesswork at this stage, the region's primary center should have been the city and bishopric of Kition, present-day Larnaca, 23 km northeast. In this context, it is worth looking at similar examples elsewhere in Cyprus.

In the neighboring region of Kalavasos-Kopetra, 12 km southwest of Kofinou, excavations have unearthed the remains of a prosperous rural settlement of 4 ha, home of 100 families, together with three churches, serving as physical and social landmarks for local residents [69]. According to the excavator, the churches and other archaeological evidence in Kalavasos reflect the economic success of this Cypriot community, identified as a 'market village' and its control of transport and exchange on a sub-regional level [69].

The second example concerns the site of Pyla-Koutsopetria, 32 km northeast of Kofinou, where a surface survey has identified an enormous coastal site of 40 ha, with plentiful ceramic evidence confirming the agricultural and quarrying profession of its inhabitants, but most importantly, their engagement in maritime trade as their main economic activity [63]. Moreover, monumentality is also present at the site. Excavations by the Department of Antiquities in the past have revealed a basilica with *opus sectile* floors, while the results of recent geophysical prospection by the University of Dakota indicate the existence of more churches at the site [63]. Thus, Pyla-Koutsopetria functioned as an *emporion* in Late Antiquity, involved in the inter-regional distribution of Cypriot goods.

It goes without saying that every respectful second-rank settlement—in our case the agro-town of Kofinou—should be the focus and local center of a series of satellite minor rural establishments, such as hamlets, villas and farmsteads. Indeed, three small loci of ceramic concentrations east of the large settlement of Kofinou, comprising mainly of roof tile, transport and storage vessels, have been interpreted as small farms, housing a number of farming families closer to their fields (Figure 8).

Farmsteads and villa estates are amongst the commonest rural sites identified in Cyprus and beyond (from Spain and Italy to the Levant) throughout Late Antiquity. Previous survey work on the island (e.g., in the territory of Kourion) has revealed that farmsteads were usually small in size (ranging between 0.01 and 0.4 ha), had access to fresh water, and were located in prominent positions overlooking the surrounding countryside and the sea [70].

The pattern that emerges when one focuses on archaeological evidence from our site dated to the Late Antique era is of particular importance here. The micro-region of Kofinou, very much as the micro-regions of Pyla-Koutsopetria and Kalavasos-Kopetra, was characterized by the presence of a main settlement with associated basilicas. Thus, in a way, the presence of one or more basilicas at these places possibly indicated the settlement's status as a local center, with churches supervising (in a way) agricultural, processing, distribution and sometimes industrial activities. These secondary places may have varied in size and function: Pyla-Koutsopetria, a port-town of 40 ha should have provided home to 800–1000 families and functioned as an *emporion* distributing goods inter-regionally; Kofinou, an agro-town of 13 ha must have accommodated up to 300 families and participated in intensive agricultural production, storage and distribution of goods to nearby cities and port-towns; Kalavasos-Kopetra, a market village of 4 ha with a population of 100 families functioned as a principal market for local products. The primary center or regional central place, towards which these second-rank settlements were oriented, is always a nearby city, usually with a bishop, such as Kition to the east of Kofinou and Amathous to the west. Finally, third-rank settlements were satellite minor farming establishments without a church, or settlements occupied seasonally by a labor force residing in cities, port-towns, agro-towns and market-villages, and commuting seasonally between secondary settlements and their farms. This settlement hierarchy for Cyprus is always adapted by SeSaLaC according to the period under investigation, and is primarily based on deterministic factors of what makes central and secondary places. It should be born in mind, however, for periods about which archaeological data or textual evidence are confined or lacking, that central-place functions might be dispersed between a variety of sites and places, while a central person might be as important as a central place [71]. It is evident that there is clearly much more going on in the case of the Xeros valley (and other fertile and well-populated regions in Cyprus) than a three-level settlement hierarchy and dots on the map, as can be illustrated by land capacity and population estimates for the period (see below).

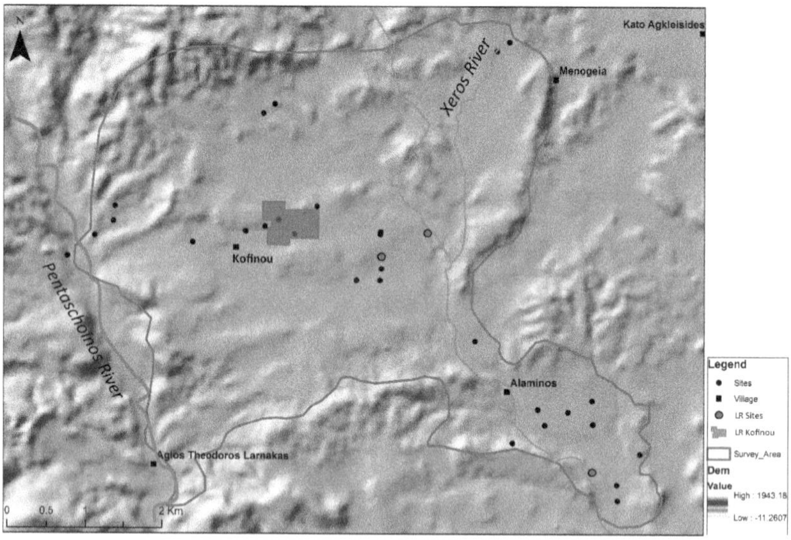

Figure 8. Late Antique (LR) sites in the Xeros valley. Image© SeSaLaC (digital geological data courtesy of the Geological Survey Department, Republic of Cyprus; GIS mapping by N. Kyriakou).

Taking Kofinou as a secondary settlement in Late Antiquity, one wonders whether the land available in its micro-region and the approximate number of people that lived in the settlement were actually compatible. Although the immediate surroundings of the Christian basilica at Kofinou nowadays give the impression of a rich and intensively cultivated landscape, the *Soil Atlas of Europe* shows that the best and most fertile soils within our survey area and the Kofinou settlement chamber make a total of 1510 ha (Figure 9). Interestingly, the Late Antique settlement itself lies in the middle of less fertile soils, a very wise choice on behalf of its inhabitants, making use of less productive areas for their settlement's built space, as well as for less demanding cultivations, such as vegetable gardens and olive groves or as pasture land. More demanding crops, such as wheat and vines, would have been cultivated in the areas with the best soils, lying 800 m away from the settlement. Considering that approximately 300 families were living in Kofinou, and taking into account that 3.6 ha of land were required per family to meet their subsistence needs in pre-industrial times [72], we arrive at the figure of 1080 ha needed for feeding the population of Kofinou. That means that the remaining 430 ha would be reserved for sustaining the population of satellite villas and farmsteads and, of course, for the export of a significant surplus that would bring in the necessary cash for the community to meet its tax obligations.

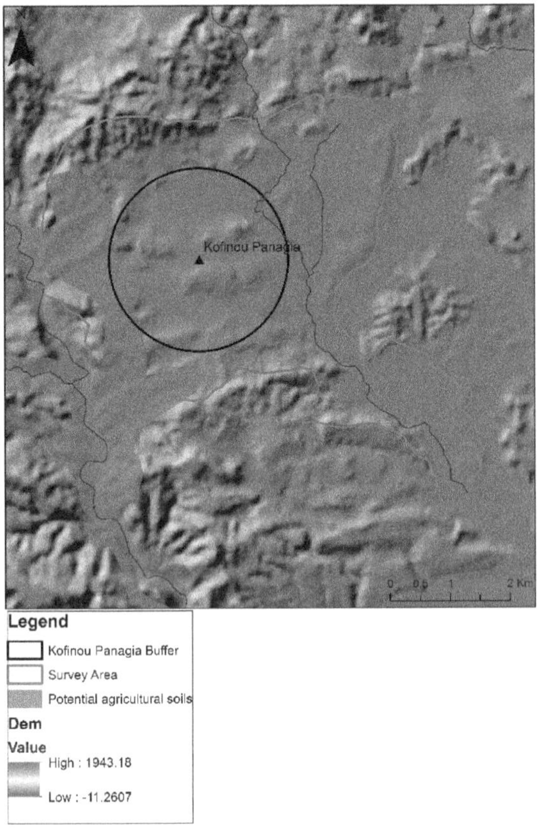

Figure 9. Best agricultural soils in the Xeros valley according to the *Soil Atlas of Europe*. Image© SeSaLaC (digital geological data courtesy of the Geological Survey Department, Republic of Cyprus; map drafted by N. Kyriakou).

It becomes evident that basilica churches mark monumental space and feature prominently within settlements of some status in Late Antique Cyprus and beyond. Apart from their role as buildings of religiousness and piety or symbols of imperial ideology and Christian identity, churches became public meeting places and focuses of production, commercial and economic activities of civic and rural communities. Examples from different regions of the Eastern Roman Empire (Byzantine Empire) are indicative of the basilica's multiple roles. In urban environments, churches began to encroach on the traditional public spaces. Such basilicas, usually of monumental dimensions, occupy part of the public area of the town, replacing in a way the Roman *forum*, and dominate the townscape, fitting into the town's layout without necessarily becoming a haphazard conglomeration [67]. The main basilica at Kourion can be a fine example of this. In peri-urban environments, extra-mural churches were usually placed along main road axes leading to the city in order to impress anyone travelling into town. In a rural context, agricultural resources seem to have been directly associated with the sighting of churches in Late Antique landscapes. Similarly to the ancient sanctuaries, Late Antique basilicas played an important role in inscribing social memory, territorial significance, and economic activities on the landscape. Last but not least, one of the roles that basilicas in rural environments seem to have assumed was that of supervisor of industrial, processing and storage activities. Excavated basilicas in Cyprus provide evidence for similar multiple roles. They appear to be encroaching civic public space, they are built close to gates and ports, in the periphery of cities and along major communication axes, but most importantly, some of them, especially urban ones, imply some kind of engagement in industry and water-management [68]. The basilica and agro-town of Kofinou may have functioned in a similar fashion: as a collection, storage center, and distribution point of agricultural produce. A link has been suggested between oil production at basilica C at Peyia, Ayia Varvara in Amathous, the northern and southern basilicas at Arsinoe and their nearby harbors. Since some basilicas with industrial and storage installations had direct access to port facilities, it seems reasonable to assume that the church had interests in the export trade, which accounts for its wealth in the sixth and early seventh centuries AD. There is no reason to believe otherwise for the case of basilicas in secondary settlements, such as *emporia*, agro-towns and market-villages. Evidence suggests that the coastal basilica at Pyla-Koutsopetria had direct access to port facilities and warehouses, while press weights and grinding stones from the vicinity of the basilicas at Kalavasos-Kopetra indicate involvement in the processing of agricultural goods [63,64]. The basilica and agro-town of Kofinou may have functioned in a similar fashion: as a collection and storage center, and distribution point of agricultural produce.

5. Concluding Remarks

This article has integrated archaeological data with landscape theory and interpretative issues. Nonetheless, a final concluding remark deriving from this macro-historic approach in the framework of the UnSaLa and SeSaLaC research projects remains to be brought to the forefront more explicitly: the sacred landscapes of Cyprus should be seen in the framework of an insular scheme relating to the transition from segmented to unitary administration (or politico-economic and ideological orientations) and vice versa. Ancient extra-urban sanctuaries acted as symbolic, territorial (even liminal in some instances when coming to the so-called frontier sanctuaries) and economic nodes within and/or between urban centers, second-rank settlements, villages and farmsteads. Similarly, Early Christian basilicas in the Cypriot countryside also acted as important symbolic and economic nodes within and/or between urban centers, second-rank agro-towns, market villages, or monastic centers in close proximity to regional central places or cities/bishoprics, and satellite farm estates. It is clear that, as in earlier periods of Cypriot antiquity, at the transition from the Roman period to Late Antiquity, we move from a half-empty to a half-full and later to a full sacred landscape. This is a much celebrated phase of Cypriot religiosity, extending well beyond the limits of monocausal explanations.

Without denying complex processes of transmission and transformation of culture, our approach remains deeply Cypro-centric and macro-historic, actively situating external influences within a local context, while acknowledging the capacity of the Cypriot insular habitual systems to accommodate and transform. This approach is particularly pertinent in the study of ancient Cypriot ritual and religion [37,73–76]. We are confident that both individual projects under the umbrella of the recently initiated UnSaLa research network, and our SeSaLaC regional surface survey project in the Xeros valley, will further add to the understanding of ancient sacred landscapes in Cyprus, and beyond.

Acknowledgments: This research falls within the umbrella of a larger collaborative network titled 'Unlocking Sacred Landscapes' (UnSaLa) (for further information see http://www.ucy.ac.cy/unsala/, accessed on 26 April 2017), previously funded by the Irish Research Council, and currently funded by the 'Research Training Group 1878: Archaeology of Pre-Modern Economies', a joint research program of the Universities of Bonn and Cologne (funded by the Deutschen Forschungsgemeinschaft (DFG)). The primary archaeological data discussed in the last part of this article come from the 'Settled and Sacred Landscapes of Cyprus' (SeSaLaC) surface survey project (http://www.ucy.ac.cy/artlands/en/research/sesalac, accessed on 2 May 2017), funded by the University of Cyprus (2014–2017), Dumbarton Oaks (2015–2016) and the Irish Research Council (2014).

Author Contributions: Both authors contributed equally to the writing of the article. This is the result of an ongoing research collaboration, long discussion and exchange of expertise between the two authors. Author names are listed in alphabetic order.

Conflicts of Interest: The authors declare no conflict of interest.

1. Swiny, S. The rôle of intuitive and small scale surveys in landscape archaeology. In *Archaeological Field Survey in Cyprus. Past History, Future Potentials: Proceedings of a Conference Held by the Archaeological Research Unit of the University of Cyprus, 1–2 December 2000*; Iacovou, M., Ed.; British School at Athens Studies 11; The British School at Athens: London, UK, 2004; pp. 55–61.
2. Finlayson, B.; Dennis, S. Landscape, archaeology and heritage. *Levant* **2002**, *34*, 219–227. [CrossRef]
3. Johnson, M. Thinking about Landscape. In *Archaeology: The Key Concepts*; Renfrew, C., Bahn, P., Eds.; Routledge: London, UK, 2005; pp. 156–159.
4. Agnew, J.A.; Duncan, J.S. (Eds.) *The Power of Place. Bringing Together Geographical and Sociological Imaginations*; Unwin Hyman: London, UK, 1989.
5. Knapp, A.B.; Ashmore, W. Archaeological landscapes: Constructed, conceptualized, ideational. In *Archaeologies of Landscape: Contemporary Perspectives*; Ashmore, W., Knapp, A.B., Eds.; Blackwell: Oxford, UK, 1999; pp. 1–30.
6. Given, M. *The Archaeology of the Colonized*; Routledge: London, UK, 2004; p. 18.
7. Wilkinson, T.J. *Archaeological Landscapes of the Near East*; The University of Arizona Press: Tuscon, AZ, USA, 2003; p. 4.
8. Ashmore, W. Social archaeologies of landscape. In *A Companion to Social Archaeology*; Meskell, L., Preucel, R.W., Eds.; Blackwell: Oxford, UK, 2004; pp. 255–266.
9. Gamble, C. *Archaeology: The Basics*; Routledge: London, UK, 2004; pp. 137–138.
10. Bender, B. (Ed.) *Landscape: Politics and Perspectives*; Berg: Oxford, UK, 1993; p. 3.
11. Plumwood, V. The concept of a cultural landscape: Nature, culture and agency in the land. *Ethics Environ.* **2006**, *11*, 115–150. [CrossRef]
12. Whatmore, S.J.; Hinchliffe, S. Ecological landscapes. In *The Oxford Handbook of Material Culture Studies*; Hicks, D., Beaudry, M.C., Eds.; Oxford University Press: Oxford, UK, 2015; pp. 1–20.
13. Thomas, J. The politics of vision and the archaeologies of landscape. In *Landscape: Politics and Perspectives*; Bender, B., Ed.; Berg: Oxford, UK, 1993; pp. 19–48.
14. Alcock, S.E. *Graecia Capta: The Landscapes of Roman Greece*; Cambridge University Press: Cambridge, UK, 1993.
15. Alcock, S.E. Minding the gap in hellenistic and roman greece. In *Placing the Gods: Sanctuaries and Sacred Space in Ancient Greece*; Alcock, S.E., Osborne, R., Eds.; Clarendon Press: Oxford, UK, 1994; pp. 247–261.
16. Alcock, S.E. The reconfiguration of memory in the eastern Roman Empire. In *Empires: Perspectives from Archaeology and History*; Alcock, S.E., D'Altroy, T.N., Morrison, K.D., Sinopoli, C.M., Eds.; Cambridge University Press: Cambridge, UK, 2001; pp. 323–350.
17. Alcock, S.E.; Gates, J.E.; Rempel, J.E. Reading the landscape: Survey archaeology and the hellenistic *Oikoumene*. In *A Companion to the Hellenistic World*; Erskine, A., Ed.; Blackwell: Oxford, UK, 2003; pp. 355–372.
18. Van Dyke, R.M.; Alcock, S.E. (Eds.) *Archaeologies of Memory*; Blackwell: Oxford, UK, 2003.
19. Tilley, C. *A Phenomenology of Landscape: Places, Paths and Monuments*; Berg: Oxford, UK, 1994.
20. Brück, J. Experiencing the past? The development of a phenomenological archaeology in British prehistory. *Archaeol. Dialogues* **2005**, *12*, 45–72.
21. Johnston, R. Approaches to the perception of landscape: Philosophy, theory, methodology. *Archaeol. Dialogues* **1998**, *5*, 54–68. [CrossRef]
22. Crumley, C.L. Sacred landscapes: Constructed and conceptualized. In *Archaeologies of Landscape: Contemporary Perspectives*; Ashmore, W., Knapp, A.B., Eds.; Blackwell: Oxford, UK, 1999; pp. 269–285.
23. Cherry, J.F. Cyprus, the Mediterranean, and survey: Current issues and future trends. In *Archaeological Field Survey in Cyprus. Past History, Future Potentials. Proceedings of a Conference held by the Archaeological Research Unit of the University of Cyprus, 1-2 December 2000*; Iacovou, M., Ed.; British School at Athens Studies 11; The British School at Athens: London, UK, 2004; pp. 23–35.
24. Smith, J.S. Preliminary comments on a rural Cypro-Archaic sanctuary in Polis-Peristeries. *Bull. Am. Sch. Orient. Res.* **1997**, *308*, 77–98. [CrossRef]
25. Beer, C. Ethnic diversity and financial differentiation in Cypriot sanctuaries. In *Economics of Cult in the Ancient Greek World: Proceedings of the Uppsala Symposium 1990*; Linders, T., Alroth Boreas, B., Eds.; Uppsala Studies in Ancient Mediterranean and Near Eastern Civilizations 21; Acta Universitatis Upsaliensis: Uppsala, Sweden, 1992; pp. 73–84.

26. Fischer, B. Le rôle des sanctuaires dans l'économie chypriote. *Cahier du Centre d'Études Chypriotes* **2001**, *31*, 51–58. [CrossRef]
27. Collombier, A.M. Organisation du territoire et pouvoirs locaux dans l'ile de Chypre à l'époque perse. *Transeuphratène* **1991**, *4*, 21–43.
28. Collombier, A.M. Quelques jalons pour une histoire de l'identité chypriote á l'époque des royaumes autonomes. In *Mediterranée: Ruptures et Continuités. Actes du Colloque tenu à Nicosie les 20–22 Octobre 2001, Université Lumière-Lyon 2, Université de Chypre*; Chehab, M., Ioannou, Y., Métral, F., Eds.; Travaux de la Maison de l'Orient Méditerranéen 37; Maison de l'Orient et de la Méditerranée: Lyon, France, 2003; pp. 139–150.
29. Masson, O.; Hermary, A. La géographie des royaumes chypriotes chez les modernes. *Cahier du Centre d'Études Chypriotes* **1992**, *17*, 23–28. [CrossRef]
30. Hermary, A. Les sanctuaires chypriotes. *L'archéologue Archéologie Nouvelle* **1996**, *25*, 38–43.
31. Hermary, A. Votive offerings in the sanctuaries of Cyprus, Rhodes and Crete during the late geometric and archaic periods. In *Proceedings of the International Symposium: Eastern Mediterranean: Cyprus-Dodecanese-Crete, 16th–6th Cent. B.C.: Rethymnon 13–16 May 1997*; Karageorghis, V., Stampolides, N.C., Eds.; University of Crete and The A.G. Leventis Foundation: Athens, Greece, 1998; pp. 265–276.
32. Hermary, A. Recherches récentes sur le territoire et les frontières des royaumes chypriotes (VIIIe –IVe siècles av. J.-C.). Proceedings of Grenzen International Kolloquium on Ritual und Kult der Antike, Basel, Switzerland, 5–6 November 2009; Guggisberg, M.A., Ed.; Schwabe Verlag: Basel, Switzerland, 2013; pp. 116–129.
33. Fourrier, S. Les territoires des royaumes Chypriotes archaïques: Une esquisse de géographie historique. *Cahier du Centre d'Études Chypriotes* **2002**, *32*, 135–146. [CrossRef]
34. Fourrier, S. *La Coroplastie Chypriote Archaïque. Identités Culturelles et Politiques à L'époque des Royaumes*; Travaux de la Maison de l'Orient et de la Méditerranée 46; Maison de l'Orient et de la Méditerranée: Lyon, France, 2007.
35. Counts, D.B. Art and religion in the Cypriote Mesaoria: The view from Athienou-*Malloura*. *Cahier du Centre d'Études Chypriotes* **2004**, *34*, 173–190. [CrossRef]
36. Papantoniou, G. Cypriot sacred landscapes from *Basileis* to *Strategos*: Methodological and interpretative approaches. In *Postgraduate Cypriot Archaeology (POCA 2005), Proceedings of the Fifth Annual Meeting of Young Researchers on Cypriot Archaeology, Dublin, Ireland, 21–22 October 2005*; Papantoniou, G., Ed.; British Archaeological Reports, International Series 1803; Archaeopress: Oxford, UK, 2008; pp. 37–45.
37. Papantoniou, G. *Religion and Social Transformations in Cyprus. From the Cypriot 'Basileis' to the Hellenistic 'Strategos'*; Mnemosyne Supplements 347; Brill: Brill, The Netherlands, 2012.
38. Papantoniou, G. Cypriot sanctuaries and religion in the Early Iron Age: Views from before and after. In *Cyprus and the Aegean in the Early Iron Age—The Legacy of Nicolas Coldstream*; Iacovou, M., Ed.; Bank of Cyprus Cultural Foundation: Nicosia, Cyprus, 2012; pp. 285–319.
39. Papantoniou, G. Cyprus from Basileis to Strategos: A sacred-landscapes approach. *Am. J. Archaeol.* **2013**, *117*, 33–57. [CrossRef]
40. Papantoniou, G. Cypriot autonomous polities at the crossroads of empire: The imprint of a transformed islandscape in the Classical and Hellenistic periods. *Bull. Am. Sch. Orient. Res.* **2013**, *307*, 169–205. [CrossRef]
41. Papantoniou, G.; Kyriakou, N.; Sarris, A.; Iacovou, M. Sacred Topography in Iron Age Cyprus: The case of Vavla-Kapsalaes. In *Archaeological Research in the Digital Age, Proceedings of the 1st Conference on Computer Applications and Quantitative Methods in Archaeology Greek Chapter (CAA-GR) Crete, Rethymno, Greece, 6–8 March 2014*; Papadopoulos, C., Paliou, E., Chrysanthi, A., Kotoula, E., Sarris, A., Eds.; Institute for Mediterranean Studies: Rethymno, Greece, 2015; pp. 70–75.
42. Ulbrich, A. *Kypris. Heiligtümer und Kulte Weiblicher Gottheiten auf Zypern in der Kyproarchaischen und kyproklassischen Epoche (Königszeit)*; Alter Orient und Altes Testament 44; Ugarit Verlag: Münster, Germany, 2008.
43. Marquaille, C. The External Image of Ptolemaic Egypt. Ph.D. Thesis, King's College, University of London, London, UK, 2001; p. 139.
44. Mitford, T.B. The character of Ptolemaic rule in Cyprus. *Aegyptus* **1953**, *33*, 80–90.
45. Bagnall, R.S. *The Administration of the Ptolemaic Possessions outside Egypt*; Brill: Leiden, The Netherlands, 1976; pp. 57–58.

46. Rautman, M. From polytheism to christianity in the temples of Cyprus. In *Ancient Journeys. A Festschrift in Honour of Eugene Numa Lane*; Callaway, C., Ed.; 2001; The STOA; Available online: http://www.stoa.org/hopper/text.jsp?doc=Stoa:text:2001.01.0014 (accessed on 4 May 2017).
47. Papantoniou, G. From segmentation to unification: Sacred landscapes and sculpture in the construction of Hellenistic Island identities. In *Hellenistic Cyprus*; Scherrer, P., Koiner, G., Ulbrich, A., Eds.; Keryx 2; Zentrum Antike Universität Graz: Graz, Austria, 2012; pp. 91–105.
48. Papantoniou, G. Rethinking Cypriot Hellenistic portraits. In *Ancient Cyprus Today: Museum Collections and New Research Approaches to the Archaeology of Cyprus*; Bourogiannis, G., Mühlenbock, C., Eds.; Studies in Mediterranean Archaeology PB 184; Åströms Förlag: Uppsala, Sweden, 2016; pp. 339–350.
49. Orser, C.E., Jr. Resistance. In *Encyclopedia of Historical Archaeology*; Orser, C.E., Jr., Ed.; Routledge: London, UK, 2002; pp. 476–477.
50. Connelly, J.B. *Votive Sculpture of Hellenistic Cyprus*; Cyprus Department of Antiquities: Nicosia, Cyprus, 1988; p. 9.
51. Senff, R. Roman sculpture in Cyprus: Imperial art and local tradition. In *Cyprus and the East Mediterranean in the Iron Age, Proceedings of the Seventh British Museum Classical Colloquium, April 1988*; Tatton-Brown, V., Ed.; British Museum Press: London, UK, 1989; pp. 188–192.
52. Poyiadji-Richter, E. Roman portraits on Cypriot grave reliefs. *Cahiers du Centre D'études Chypriotes* **2009**, *39*, 177–196. [CrossRef]
53. Papantoniou, G. 'Hellenising' the 'Cypriot Goddess': 'Reading' the Amathousian terracotta figurines. In *From Pella to Gandhara. Hybridisation and Identity in the Art and Architecture of the Hellenistic East*; Kouremenos, A., Chandrasekaran, S., Rossi, R., Eds.; British Archaeological Reports International Series 2221; Archaeopress: Oxford, UK, 2011; pp. 35–48.
54. Morris, C.E.; Papantoniou, G. Cypriot-born Aphrodite. The Social biography of a modern cultural icon. In *Il Trono Variopinto. Figure e Forme della dea Dell'amore*; Bombardieri, L., Braccini, T., Romani, S., Eds.; Hellenica, Testi e Strumenta di Letteratura Greca Antica, Mediavale e Umanistica 55; Edizioni Universitarie dell' Orso: Torino, Italy, 2014; pp. 183–202.
55. Mitford, T.B. The Hellenistic inscriptions of Old Paphos. *Ann. Br. Sch. Athens* **1961**, *56*, 1–41. [CrossRef]
56. Potter, D. H Κύπρος Επαρχία της Ρωμαϊκής Αυτοκρατορίας. In *Ιστορία της Κύπρου*; Volume 2, *Αρχαία Κύπρος*. Part 2; Papadopoullos, T., Ed.; Archbishop Makarios III Foundation: Nicosia, Cyprus, 2000; pp. 763–864.
57. Hellmann, M.C.; Hermary, A. Inscriptions d'Amathonte, III. *Bull. Corresp. Hell.* **1980**, *104.1*, 259–275.
58. Hermary, A.; Schmid, M. Le sanctuaire d'Aphrodite. *Bull. Corresp. Hell.* **1987**, *111*, 735–740.
59. Fourrier, S.; Hermary, A. (Eds.) *Amathonte VI: Le Sanctuaire D'aphrodite des Origines au Début de L'époque Impériale*; Études Chypriotes 17; École Française d'Athènes: Athens, Greece, 2006; pp. 163–164.
60. Papantoniou, G. 'Revisiting' Soloi-Cholades: Ptolemaic power, religion, and ideology. *Cahier du Centre d'Études Chypriotes* **2009**, *39*, 271–287. [CrossRef]
61. Given, M.; Knapp, A.B. *The Sydney Cyprus Survey Project. Social Approaches to Regional Archaeological Survey*; Monumenta Archaeologica Series 21; Cotsen Institute of Archaeology, University of California: Los Angeles, CA, USA, 2003; p. 313.
62. Chrysos, E. Cyprus in Early Byzantine Times. In *The Sweet Land of Cyprus: Papers Given at the Twenty-Fifth Jubilee Spring Symposium of Byzantine Studies, Birmingham, March, 1991*; Bryer, A.A.M., Georghallides, G.S., Eds.; Cyprus Research Centre: Nicosia, Cyprus, 1993; pp. 3–14.
63. Caraher, W.R.; Moore, S.; Pettegrew, D.K. *Pyla-Koutsopetria I. Archaeological Survey of an Ancient Coastal Town*; American School of Oriental Research Archaeological Reports 21; American School of Oriental Research: Boston, MA, USA, 2014.
64. Rautman, M. The Troodos in Twilight: A provincial landscape in the time of Justinian. In *Cyprus and the Balance of Empires: Art and Archaeology from Justinian I to the Coeur de Lion*; Davis, T., Stewart, C.A., Weyl Carr, A., Eds.; American Schools of Oriental Research Archaeological Reports 20; American School of Oriental Research: Boston, MA, USA, 2014; pp. 39–56.
65. Kyriakou, N. Sacred landscapes in late Roman Cyprus: Cityscapes and peripheries in context. In *Spatial Analysis of Ritual and Cult in the Mediterranean*; Papantoniou, G., Morris, C., Vionis, A.K., Eds.; Studies in Mediterranean Archaeology; Åströms Förlag: Uppsala, Sweden, under review.

66. Rautman, M. *A Cypriot Village of Late Antiquity: Kalavasos–Kopetra in the Vasilikos Valley*; Journal of Roman Archaeology Supplementary Series 42; J.H. Humphrey: Portsmouth, UK, 2003; pp. 241–242.
67. Vionis, A.K. Settled and sacred landscapes of Cyprus: The relationship of church and landscape in Byzantine Kofinou. In *Church Building in Cyprus (4th to 7th Centuries): A Mirror of Intercultural Contacts in the Eastern Mediterranean*; Rogge, S., Nicolaou, D., Horster, M., Eds.; Schriften des Instituts für Interdiziplinäre Zypern-Studien; Waxmann: Münster, Germany, 2016.
68. Vionis, A.K.; Papantoniou, G. Sacred landscapes as economic central places in Late Antique Naxos and Cyprus. *Antiquité Tardive: Revue Internationale d'Histoire et d'Archéologie (IVe-VIIe Siècle)*, under review.
69. Rautman, M. The villages of Byzantine Cyprus. In *Les Villages dans L'empire Byzantine*; Lefort, J., Morrisson, C., Sodini, J.-P., Eds.; Lethielleux: Paris, France, 2006; pp. 453–463.
70. Mavromatis, C.; Swiny, S. Land behind Kourion: Results of the 1997 Sotira Archaeological Project Survey. *Report of the Department of Antiquities, Cyprus*, 2000; 433–452.
71. Vionis, A.K. Understanding settlements in Byzantine Greece: New data and approaches for Boeotia, sixth to thirteenth centuries. *Dumbart. Oaks Pap.* **2017**, *71*. in press.
72. Shiel, R.S.; Stewart, A. The soils and agricultural potential of the Thespiai area. In *Testing the Hinterland: The Work of the Boeotia Survey (1989–1991) in the Southern Approaches to the City of Thespiae*; Bintliff, J.L., Howard, P., Snodgrass, A., Eds.; McDonald Institute Monographs: Cambridge, UK, 2007; pp. 95–109.
73. Papantoniou, G. The 'Cypriot Goddess' at the transition from the Bronze to the Iron Age: A 'Cypro-centric' approach. In *J.R.B. Stewart. An Archaeological Legacy*; Knapp, A.B., Webb, J.M., McCarthy, A., Eds.; Studies in Mediterranean Archaeology 139; Åströms Forlag: Uppsala, Sweden, 2013; pp. 161–173.
74. Papantoniou, G. Cypriot ritual and cult from the Bronze to the Iron Age: A *longue-durée* approach. *J. Greek Archaeol.* **2016**, *1*, 73–108.
75. Iacovou, M. Advocating Cyprocentricism: An Indigenous model for the emergence of state formation on Cyprus. In *"Up to the Gates of Ekron". Essays on the Archaeology and History of the Eastern Mediterranean in Honor of Seymour Gitin*; White Crawford, S., Ben-Tor, A., Dessel, J.P., Dever, W.G., Mazar, A., Aviram, J., Eds.; The W.F. Albright Institute of Archaeological Research, The Israel Exploration Society: Jerusalem, Israel, 2007; pp. 461–475.
76. Counts, D.B.; Iacovou, M. New approaches to the elusive Iron Age polities of ancient Cyprus: An introduction. *Bull. Am. Sch. Orient. Res.* **2013**, *370*, 1–13. [CrossRef]

land

MDPI

Article

Using Historical Maps within a GIS to Analyze Two Centuries of Rural Landscape Changes in Southern Italy

Dina Statuto *, Giuseppe Cillis and Pietro Picuno

SAFE School of Agricultural, Forest, Food and Environmental Sciences, University of Basilicata, Viale Ateneo Lucano, 10, 85100 Potenza, Italy; giucillis@gmail.com (G.C.); pietro.picuno@unibas.it (P.P.)
* Correspondence: dina.statuto@unibas.it; Tel.: +39-3298-439319

Received: 30 July 2017; Accepted: 11 September 2017; Published: 19 September 2017

Abstract: The current characteristics of a rural landscape may be better understood if suitable information related to its past is available. The availability of a Geographical Information System (GIS) can enable the analysis of landscape features in relation to several aspects, e.g., the evolution and mutual inter-relation among different ecosystems, the impact and sustainability of human activities, the visual characteristics of a landscape, etc. The analysis of geographical information, derived from historical maps, within a GIS could, therefore, prove to be a very powerful tool, for a better-informed decision-making and management of a rural landscape. With the aim to identify the land use changes in a rural area located in the Basilicata Region (Southern Italy), a territorial analysis was conducted through a GIS, in which data taken from historical maps—covering a period of 184 years, from 1829 to 2013—were implemented. Three-dimensional reconstruction of the rural landscape during different periods were obtained through digital terrain models (DTM). The land cover changes were also evaluated, demonstrating how they have affected the quality of the forest ecosystem in the area. The final results that were obtained comparing historical documents and current maps enabled the evaluation of the multi-temporal, morphological, and vegetation variations in this rural landscape. The analysis that was conducted has great potential for assessing and monitoring landscape diversity and typical changes of vegetation, even in different geographical locations, where appropriate interventions in landscape structures may be so planned.

Keywords: geographical information system; historical maps; landscape changes; rural land

1. Introduction

During the last decades, changes in land use and management have led to the degradation of several cultural landscapes in rural areas, mostly in those located in the Eastern Mediterranean, with relevant consequences for local populations, landscape functionality, and the maintenance of ecosystem services [1]. The analysis of rural land modifications, as well as the wider environment and landscape context in which they take place, is important in order to understand the profound transformations connected with human intervention and natural events [2,3].

The digitization of historical cartography allows for understanding the evolution of the landscape over time, and may account for the susceptibility of spaces to territorial transformations and provide insights to spatial and contextual specifics, which is an issue that is essential for appropriate spatial planning [4]. With the support of a GIS in which historical documents are input, a multi-temporal analysis of land is possible. This is a fundamental tool for monitoring landscape diversity and changes in vegetation and landscape structure [5–7].

Visibility analysis may serve as the starting point to evaluate the quality, fragility, and aptitude for the protection of the area's biophysical units. This analysis has a two-fold importance:

i.e., for planning purpose since it can be used to identify which biophysical units are in greater need of protection and where such protection is more likely to be effective, and because it provides a novel methodological basis for the analysis of visual aspects in biophysical units [8,9]. During the past two decades, especially in the Mediterranean Basin, three phenomena are re-drawing the configuration of rural areas: agricultural mechanization, the accelerating demise of traditional rural life, and an increase in the mobility of individuals [10]. In fact, in many developed countries, a particular land-cover change pattern has taken place: plains are being increasingly utilized for human activities, while mountain areas are being abandoned, so they are undergoing natural reforestation processes [11–16]. These generalized patterns can have significant impacts on biodiversity distribution and conservation [17].

Image processing techniques and landscape pattern metrics were applied to quantify the changes in forest cover patterns, while appropriate statistical descriptors were adopted to investigate the relationship between land-cover changes and topographical factors. New systems for the rational collection and analysis of forestry and agricultural land data are now available. GIS-based techniques, image processing, remote sensing, and other powerful technologies for the survey, planning, and management of land evolution are enabling a more accurate analysis of rural landscape and environment [18]. In particular, GIS are excellent tools for landscape modelling and three-dimensional analysis. They allow an easy digitization of geographical and non-geographical information, and facilitate visual representation. A GIS approach for territorial analysis, comparing historical maps with contemporary maps, may reveal essential clues for understanding landscape dynamics [2,4,10,19].

This paper presents a local study on the land use evolution using historical and recent maps. It is aimed to analyse, in a specific way, the land use dynamics and topographic changes from 1829 to 2013 by a comparative examination of different historical cartographic supports and more recent maps. This approach would enable to obtain conclusions about the rural landscape changes, their connection to human activities—agricultural production, in particular—as well as to natural events, and the consequences on the rural land. This analysis, by proposing a methodology supported in GIS tools to evaluate changes that make comparable maps from different dates and data sources, may, therefore, enable policy decisions that support sustainable development, acknowledging present and future development trends and their potential economic, social, and ecological impacts. The technical and spatial analysis methodologies applied could ensure both the proper management and planning of land [20].

2. Materials and Methods

2.1. Study Area

The study area (Figure 1) is located in the Basilicata region (Southern Italy). It spans about 18 km^2, covering one part of the Forenza municipality (40°47′57″ N 15°51′39.4″ E, datum WGS84). It is about 5 km away from the urban centre, a hilly area often characterized by high slopes, forests and other semi-natural areas. The study area is part of the hydrographic basin of the Bradano River, which crosses the area from the northwest to the southeast for 6.5 km; after 2 km it flows into the Acerenza's dam, then continuing its course towards the Ionian Sea. The altitude of the area ranges from 450 to 920 m above sea level (a.s.l.). The soil composition has, in turn, determined the topography of this area, which has influenced the socio-economic context through relevant agricultural activities. Considering the thermo-pluviometric data of the weather station located in Forenza, the annual rainfall is 660 mm in average (Hydrological Annals of Ministry), distributed into 86 rainy days with higher values during autumn and winter. The mean annual temperature is 12.8 °C, the average monthly temperature is lowest during January with 3.4 °C, whilst the hottest month is August with a monthly average of 22.8 °C.

The territory offers employment mainly in agricultural activities, many farms present in this area have livestock and cereal crops. The largest profit comes from cattle breeding and in particular the production and sale of dairy products. Moreover, fodder, olive groves, orchards and vineyards cover the hilly territory.

The wooded area is located in the San Giuliano area. The forest communities encountered are typical of the phytoclimatic area *Castanetum* [21] and *Quercus-Tilia-Acer*-mixed deciduous forest belt of Schmid. At lower altitudes, the vegetation characteristics are similar to those of the Mediterranean basin, while those at higher altitudes of the mountain are typical of the sub-mountain belt. With nearly 4000 ha of wood, Forenza is one of the municipalities with the largest wooded area in the whole Basilicata region.

Figure 1. Location of the study area, coordinates 4517330 N 572110 E, EPSG: 32633-WGS 84/UTM zone 33N.

2.2. Cartography

Land use change and geographical reconstruction of landscape were carried-out over a period of 184 years in four time steps: years 1829, 1876, 1955, and 2013. The specific cartography for each period was examined with the aim to create different base maps to make a comparison in time. The historical maps and aerial-photography were scanned first, and then imported and georeferenced into a GIS system.

The historical map from 1829 is stored in the State Archive of Potenza: this map was produced in 1829 by legal experts to solve the division of domains after border disputes. Similar maps were realized in the whole Southern Italy as a result of the abolition of feudalism. These maps were drawn using legal documents and diplomatic sources, which showed land use rights and surveying on the ground (Figure 2). This historical map was manually drawn (577×440 mm^2) and represents a part of the Forenza municipality (San Giuliano's Wood). It shows, in yellow, the presence of agricultural land around the farms and the name of some farm districts. The evaluation through chromatic differences was possible thanks to the comparison with other similar historical maps (same period and surrounding areas), which had a more detailed description. In addition, from historical documents, some useful information were obtained, linked to the production of the map. They can be considered as ancillary data that can better describe the historical situation (Forenza Municipality, 1915).

The historical topographic map of 1876—also stored in the same State Archive of Potenza—was produced by the Italian Geographical Military Institute (IGMI) to create a complete mapping of the Italian territory after its national unification. Geodetic surveying and topographic evaluation were highly accurate, and the IGMI decided to immediately print some drafts using a photolithography technique (Figure 3). This cartography consists of 174 sheets, the study area being located into Sheet no. 65—Avigliano, eastern part. The scale of the map is 1:50,000 and the altitude was represented with contour lines having an equidistance of 10 m, and elevation of the highest areas. The old maps are a source of important information, even if they require an accurate analysis to evaluate the topographic accuracy, chronometric accuracy, and planimetric completeness [22]. The first aspect indicates the quantity and quality of information about landscape objects. Each map contains only a selection of geographic features, which are symbolized by the cartographer according to the objective of the map. Then it is necessary to consider whether the map depicts all features of a certain class, and how accurately the cartographer classified the features thematically [23]. The second aspect refers to the dating of the map as a physical artefact (by watermark analyses or other techniques), and the dating of the information contained in the map. Dating the age of map information is often difficult, as the production or the revision of a map commonly takes several years. The third aspect is the planimetric completeness (or geometric accuracy) [24].

An aero-photogrammetric survey with black and white film for almost the whole Italian territory was carried out after the Second World War, in the period between 1954 and 1955 by the Italian Aeronautics Group, on behalf of the Italian Geographic Military Institute (IGMI) and the USA Army Map Service. For the coverage of this study, two photograms were used. Each of these photos has an overlapped portion and was shot in May with a Fairchild camera (focal length of 152 mm), the size of the photo is 230 × 230 mm, the flight altitude is around 6000 m, its approximate scale is 1:33,000. The quality of the survey allows an accurate geolocation and assessment of land use. The aerial photo from 1955 is stored in the archive of the School of Agriculture, Forestry, Food, and Environmental Sciences (SAFE) of the University of Basilicata, Italy.

The recent cartographic support of the year 2013 were downloaded in GeoTIFF (.tif) format from the Geographic Database of the Basilicata Region (under Italian Open Data License 2.0 (IODL 2.0)). Orthophotos mosaic in 1:5000 scale obtained from aerial photographs taken in the year 2013 (from April to September) were used to determine land use for the year 2013. Each orthophoto is made of four RGB visible bands or near infrared with an average pixel value of 0.2 m. It is geo-referenced in the UTM/WGS84 projection. These maps are under the IODL 2.0 license, which allows free use. They combine the image characteristics of a photograph with the geometric qualities of a map, i.e., the ground features are displayed in their true position, so it is possible to carry out direct measurement of distance, areas, angles, and positions [25].

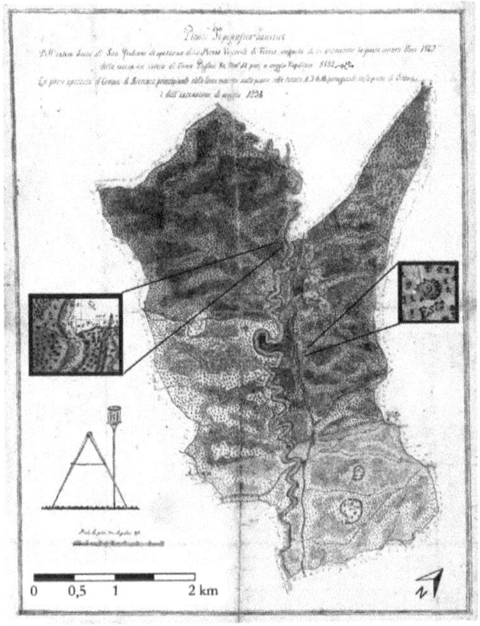

Figure 2. Historical map of the Municipality of Forenza (1829).

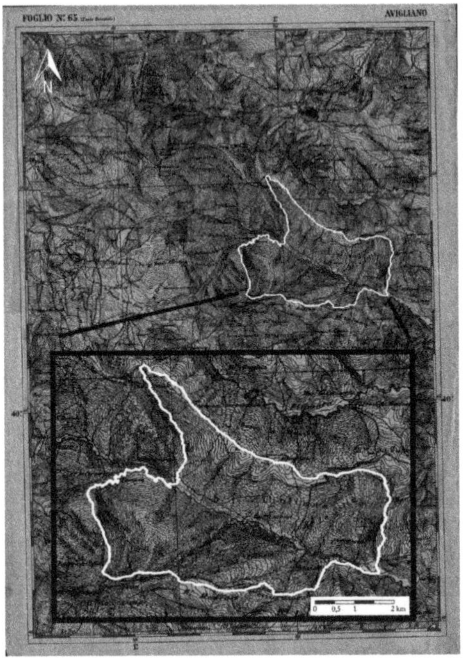

Figure 3. Historical map of the Municipality of Forenza (1876).

2.3. *Historical Documents*

Historical records provide land managers with information that could be used to understand trajectories of land use change. According to Italian and foreign studies, the related historical documents mainly include chronicles, timber ledgers, tax records, census data, land survey records forest inventories, paleoecological records, government report, gazetteer, notes of surveyors, statistical inventories, genealogy, general description by travellers, paintings, farmers' diaries, cultural histories, etc. [26]. The information reported in some historical documents may also support map reading, if there is not a legend, and explain some qualitative and quantitative data about historical land use. For this study, the historical documents analysed providing information on the main agricultural activities, forestry, and on the landscape and the environment; so it was possible to trace their evolution. Integrating a GIS methodology to quantify land use change with historical analyses makes it possible to identify the dynamics that led to the present landscaping situation. Some documents showing the relevant information for this study are reported: statistics of the Kingdom of Naples [27], a report on the ousting of St. Julian's domain forest [28], historical documents from the Forenza municipality archive. Until the early 1900s forest management was not correct and illegal logging of public domains were common. The most frequent forest logging system identified was clear-cut with reserves of extensive areas. Documents reported significant logging and tillage into the forest, which was commissioned by the Forenza municipality to solve its financial problems.

2.4. *Data Analysis*

With the aim to analyse the evolution of land use in different time step (Figure 4), all maps have been converted into data, and each homogeneous area has been identified according to its relevant category of land use. The number and features of each map depend on its cartographic properties.

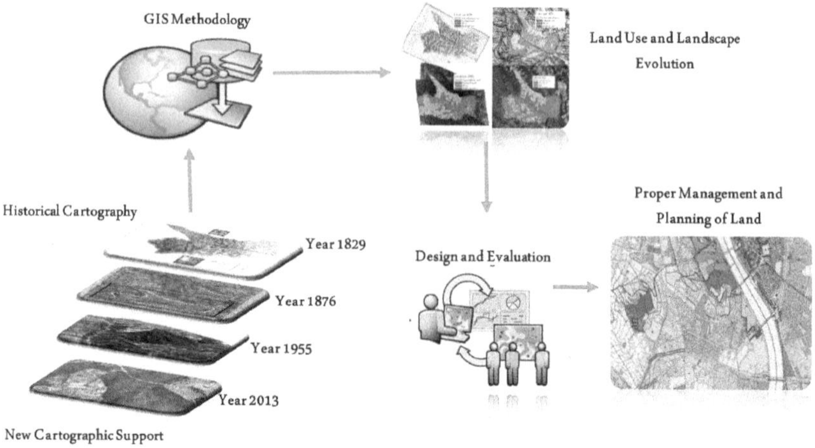

Figure 4. Applied methodology.

2.4.1. Georeferencing and Data Processing

According with Lefreniere and Rivet [29], working with historical maps is necessary in order to include many different aspects and problems that occurred over time. The first step has been to digitize the maps with a scan. Each map has been scanned into a TIFF file at a resolution of 300 dpi. These raster data have been imported in an open-source desktop software for geographic information system (QGIS 2.18.5 Las Palmas de Gran Canaria) to elaborate them. Every different temporal map was projected in the same coordinate system (EPSG: 32633-WGS 84/UTM zone 33N). To rectify historical maps (1829–1876) and aerial photos (1955), a sufficient number of control points on a geo-referenced map was identified (topographic maps of IGMI or orthophoto of year 2013). Today, a variety of transformation techniques are available, while different methods to compute and visualize distortion are possible [30,31]. In the present study, first- and third-order polynomial transformations were used. Five control points and a first-order polynomial transformation were used. Subsequently, a real geo-referencing has been performed in which seventeen control points—based on the analysis of the river, watersheds and roads—were added and a third-order polynomial transformation was used. The resulting ortho-rectification, after error checking, can be considered acceptable. Thanks to this data processing, it has been possible to make all maps comparable in terms of scale and land use classification and it has been possible to overlay them, in order to asses land use changes.

Planimetric and geodetic accuracy were evaluated and visualized with MapAnalyst, an open source application [24]. All raster maps were converted into vector data by visual interpretation and subsequent manual classification [32] and each homogeneous part of the area was identified with different categories of land use. The land use of the historical maps of 1829 was classified into five categories thanks to the interpretation and the integration with historical documents. The same categories have been used to digitize land use on the map of 1876.

Corine Land Cover (CLC) nomenclature guidelines [33] were used to assess land use. In order to have a more complete view of current land use, it has been carried out fieldwork was carried out to identify the types of land use and locate them through a GPS. The evolution of cartographic information in time has generated a progressive increase in the number of details and, therefore, the number of land use categories identified. Since the aim of this research was focused on the comparison between general categories of land use—differentiating natural areas (natural land, water lines) from those susceptible to human activities (agriculture, buildings, roads)—similar categories of land use in different time levels were aggregated into more general classes, so as to better compare homogeneous changes.

The most important categories identified in the study area have, therefore, been grouped according to the following five classes:

- *natural land*: includes forest (according to FAO definition) and transitional woodland-shrubs, an afforested land with conifers realized in the 1960s, transitional woodland-shrubs area (containing shrub or herbaceous vegetation with scattered trees and vegetation formations that may result from the degradation of the forest or a regeneration of the same for re-colonisation of non-forest areas);
- *agricultural land*: includes all types of cultivated areas, i.e., arable lands, vineyards, olive groves, permanent crops like orchards, permanent and natural pastures, natural pastures with spontaneous herbaceous vegetation, etc.;
- *urbanized area*: includes new and old farms, buildings and artificial surfaces next to farms. This category also includes infrastructure to regulate water flow;
- *road network:* includes only roads involving a significant surface and with an important traffic flow. Provincial and municipal roads are included, whilst all typologies of minor rural roads are not included; and
- *river:* includes the area corresponding to the bed of the Bradano river and the vegetation present along the river.

For each category, the total area expressed in hectares (ha) and the percentage of its variation over the time was calculated. The road network was also evaluated, the length calculated in meters (m).

2.4.2. DTM and Three-Dimensional Modelling

Through the digital processing of the maps it was possible to reconstruct the three-dimensional shape of the territory, thanks to a photo-mosaic procedure and virtual reconstructions of the land in different periods. Recently, 3D visualization has been receiving more attention as a useful tool to understand engineering phenomena, or to detect important elements that cannot be found in usual simulations.

Visualization in a virtual environment is a useful method that allows people to appreciate archaeological or historical objects through the computer [34].

2.4.3. Methods for Map Comparison and Correlation Analysis

Thanks to spatial analysis functions, it was possible to create a digital terrain model for the area. Overlapping photographic images, topographic maps and other cartographic data allowed the realization of a three-dimensional view, useful to appreciate landscape modification from different viewpoints. To create 3D reconstructions, a QGIS plugin (*Qgis2threejs*) was used. Moreover, through spatial analysis and image processing, using aerial photos from 1955, a historical reconstruction of 1829 was obtained, with a virtual jump of 184 years back. The analysis that was performed over the three-dimension images of the territory has been revealed to be a powerful mean of interpretation, since it allows some dynamic effects to be created in a virtual reality system, in which the operator has the opportunity to navigate as if he/she is walking or flying over the area at that time. Thus, the operator is able to visit—with a virtual jump backwards in time enabled by the simulation of an ante litteram flight—real-life scenarios, that would be otherwise difficult to imagine and/or reach. By doing so, all the aspects that characterized an area can be examined in suitable detail and completeness, starting from the analysis of topography and land cover and leading to anthropic components (buildings, roads, and railways, hydraulic infrastructure, such as dams, aqueducts, etc.).

By doing so, it is possible to appreciate qualitatively, in terms of land use and landform variations, the agro-forestry landscape changes, starting from a comparison between three-dimensional reconstructions of the area during different time layers. This virtual orthophoto of 1829 was realized through QGIS and free image processing software (GIMP—GNU Image Manipulation Program) which allows the preservation of geo-referencing.

3. Results

From the overlay of the different base maps, thanks to GIS technologies and the comparison between historical cartographic data and current situation, different categories of land use were identified.

3.1. *Land Use*

In 2013, the study area consisted of 65% natural area, 32.9% agricultural land, 1.1% river, 0.9% road network, and 0.1% built up area. The percentage of natural land decreased less intensively from the year 1829 to 1876, with a reduction of 100 ha (−5.4%), but more intensively from 1876 to 1955, with a reduction of 222 ha (−12.3%); agricultural land conversely increased from about 31% in 1829 to about 50% in 1955 (Table 1).

This trend reversed from 1995 to 2013: the analysis shows in fact a reduction of about 200 ha of the agricultural surface, almost counterbalanced by an increase of natural land from 48.6% to 65%. The main difference occurred highlights a mutual exchange between a reduction of the areas for agriculture and crops that gave more space to the natural area [25]. This is as a widespread trend detected in other different areas of the Basilicata region. The territorial extension of rivers decreased from about 41 ha in 1829 to about 20 ha in 2013 (Figures 5 and 6). In particular, analysing each category without aggregation, woodland reduced from about 1206 ha in 1829 to 856 ha in 2013 with a minimum peak in the year 1955 (752 ha).

In year 1829, the pastures represented 98% of agricultural land, which covered more than 31% of the study area; on the contrary, from the year 1955 to 2013 the arable land (cereals and fodder crops) constituted the 93.6% and 88.9% of agricultural land. In 2013 the territory was also covered by olive groves and vineyards (about 23 ha and 15 ha). The introduction of these crops starting from the second half of the last centuries has visually impacted the landscape; human activities and agricultural practices have, thus, influenced the landscape structure, creating a mosaic of different land use categories with relevant consequences on the ecosystems of this area.

Table 1. Analysis of land use categories in percentage (%) and surface (ha).

Land Use Categories	Years							
	1829		1876		1955		2013	
	ha	%	ha	%	ha	%	ha	%
Natural Land	1206.15	66.3	1106.55	60.9	883.94	48.6	1182.51	65.0
Agricultural land	570.3	31.4	673.60	37.0	907.84	49.9	598.04	32.9
River	40.9	2.2	36.90	2.0	25.30	1.4	19.96	1.1
Built up area	0.8	0.0	1.10	0.1	1.07	0.1	17.64	1.0
Total	**1818.15**	**100**	**1818.15**	**100**	**1818.15**	**100**	**1818.15**	**100**

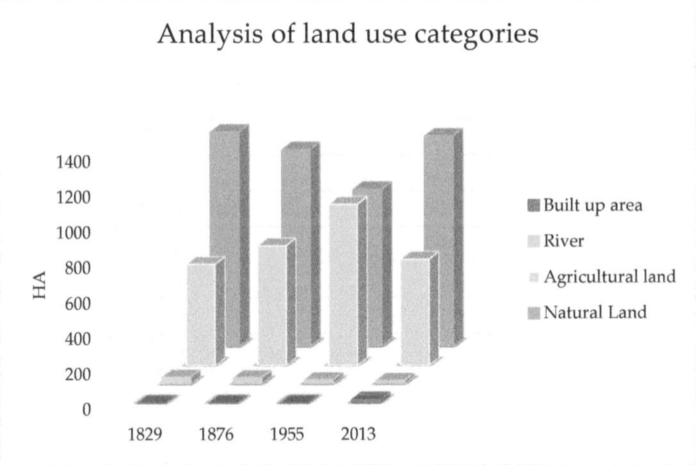

Figure 5. Analysis of land use categories in different periods.

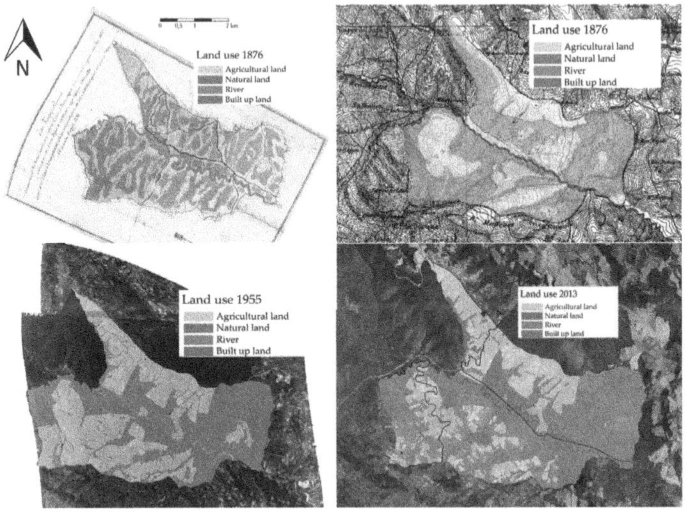

Figure 6. Spatial analysis of land use categories in different periods.

3.2. *Infrastructure*

The infrastructure was evaluated in terms of road network and farm buildings. The surface extension of the road network was estimated only in 2013; other historical maps did not allow the possibility to appropriately estimate this surface, since all roads within this study area were less than 6 m wide, so their width was not reported. Only the linear length of the road network has, therefore, been evaluated along the four analysed time steps. In the year 2013, about 0.9% of study area was crossed by roads (provincial and municipal roads that connect neighbouring towns) with a significant traffic flow. The total lengths of the road network has increased over the years; it passed from about 22 km in the year 1829 to 69 km in 2013, exclusively related to agricultural activities (paths and bridleways) until 1955 (Table 2). In 2013 about 20% of the roads were asphalted.

Farm buildings play a central role in the environmental characteristics of agricultural land over the centuries in many Eastern Mediterranean countries, since they have accompanied the development of the agricultural activities [35]. In the study area there has been an increase in their number and surface. In 1829 there were n. 13 buildings, occupying a total area of 0.8 ha, whereas in 2013 they were n = 55 and occupied an area of 2.1 ha, this value including many abandoned farms. Considering a field survey of the different buildings, it was observed that only 11 of the total 55 buildings are currently used as farms.

Table 2. Variations of the linear extension of the road network and of building number.

	Categories	
Years	**Road Network (meters)**	**Building Number (n)**
1829	22,736	13
1876	29,162	18
1955	63,862	38
2013	69,129	55

3.3. *Landscape*

Land use maps and DTM were adequately treated within the GIS to create, through spatial analysis procedures, suitable three-dimensional views, which may be used to appreciate the morphological and visual landscape variations. The overlapping of different thematic maps provides an additional database, by which a comparison of the visual aspect of the landscape with time is possible (Figure 7).

Figure 7. Three-dimensional visualization of the landscape.

Starting from the DTM of years 1876 and 1955, a slope map was created (in percentage). This type of map has been used to correlate agricultural activities with the topography of the study area and, therefore, to identify the reasons that have influenced the landscape variation over time. This type of analysis shows that about 90% of the arable land occupied a surface with a slope less than

25% (moderately steep class) in 1876. Instead, in the 1955 this value decreased to 59%, and then increased again to 78% in 2013. Figure 6 shows a hypothetical virtual orthophoto, obtained with a cross-comparison between the land use categories in the year 1955 and in 2013; it is realized by using photo-mosaic and photo-editing processes. Finally, for each year, all cartographic support and the thematic maps have been overlaid on DTM to allow a three-dimensional reconstruction of the landscape over the years. In Figure 8 it is possible to notice the land transition from agricultural activities into natural areas in the period from 1955 to 2013.

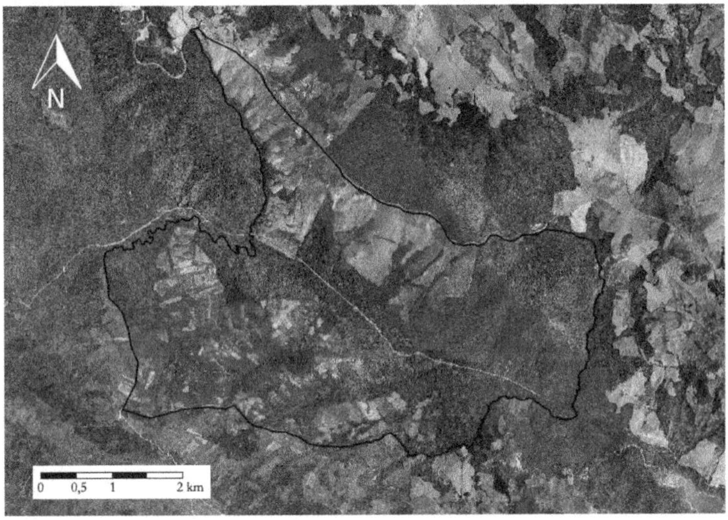

Figure 8. In red, the land transitions from agricultural activities to natural areas in the period from 1955 to 2013.

4. Discussion

The analysis showed that in 1820 the landscape of the study area was mainly dominated by oak woods, with areas with a lower forest density canopy, where grazing was the predominant agricultural activity of Forenza municipality. The importance of this activity, in particular of sheep livestock, was reported in historical documents as well as in thematic cartography (food products map of the *Kingdom of Two Sicilie's* continental provinces, 1859). Moreover, arable land was restricted around farms. The road network was not very extensive, since it was limited to a network of paths used for the connection between the neighbouring municipalities and farms. In 1829, the river was an important element for agriculture and livestock because some roads were parallel to its course, and its extension was higher than other years. After 47 years, during 1876, there was an increase of agricultural land due to higher food demand from the Forenza municipality population, which was gradually increasing. Arable lands were only extended nearby the main road network, where there were flatter and easily accessible areas.

Woodlands which, in addition, including natural meadows and transitional areas, still covered the most part of the study area at the time, were starting to become a source of income for the population, which carried out some intensive clear-cut (thanks to Bourbon forestry law that allowed deforestation with the release of 60 seed trees per hectare) aimed to solve economic problems.

After the Second World War, the landscape was significantly modified thanks to a general improvement of economic conditions and due to specific reforms targeted at the south of Italy, which resulted in radical changes in land management. In fact, land use was heavily influenced by the "*Land reform*", having among its objectives the qualitative and quantitative improvement in land use and equity for farmers [36]. The arrival of extensive agriculture, the improvement of cultivation techniques, and public financing involved a change in the landscape structure, because marginal and steep areas were then cultivated. Consequently, with the increase of agricultural activities, roads and buildings have almost doubled. The current trend, as demonstrated by the analysis of land use in the 2013, showed a progressive decrease in the agricultural areas to the benefit of natural vegetation, that growth over the years, having spontaneously covered areas that, in the past, were cultivated. This phenomenon has steadily increased over the past decades, i.e., with the development of the so-called "*set-aside*" EU regime (EEC Regulation 1272/88) by the Common Agricultural Policy (CAP) aimed to control the overproduction of cereals and other crops, in order to avoid an excessive reduction in agricultural prices [37]. Agricultural land use change is often one of the causes, but it is not always possible to identify proximate causes for the agricultural change itself, except for urbanization and the development of natural areas [38].

The analysis has allowed an assessment of the land use evolution, on landscape changes and environmental modifications over the 184 years that were investigated in terms of vegetation, environmental, and visual impact. In 1829, the landscape was dominated by oak wood in which, especially in the area near the farms and the river, the cattle could graze. From an environmental viewpoint, pastures did not cause significant problems, since they were not intensive (i.e., wooded pastures), involving about 29% of the study area. Therefore, the forested areas were the main component of the landscape, more important than in the years 1955 and 2013, in which woodland covered, respectively, only about 41% and 47% of the study area. The environmental and landscape quality also are demonstrated by the low impact that human activities had on this area: low-intensity agricultural land use, agricultural activities more suitable within the local territorial context, as well as a limited areal extent affected by rural settlements.

The increase of the population's needs, from an economic point of view, triggered landscape changes, and the study area was transformed from a (semi-)natural zone to an agricultural area. From 1876 to 1955, this over-exploitation of the forest resources has led to negative impacts on the landscape and the environment, with a significant change of its features. Using three-dimensional modelling, it has been possible to verify that land use changes and deforestation had resulted in significant modifications in the visual aspect of the rural landscape, since the forested areas of 1829 were partly replaced by arable lands (Figure 7).

Currently the area is covered by a conifer plantation (40 ha) planted during the 1960s to protect the soil, to reinforce the slopes, and to reduce the water runoff in mountainous and hilly areas [39].

A spatial overlay of the various areas, and in particular woodland, has made it possible to identify the areas that during the time period examined (i.e., 184 years) have never changed. Some degree of natural succession of vegetation has taken place depending on the time of abandonment. Species have reached considerable size and there has been a large build-up of biomass. In the last few years a decline in the vernacular farm buildings ("*masserie*") has been observed due to the abandonment of agriculture and the changes in their character. In fact, small producers have disappeared (single families) and large owners living far from that area have bought many cultivated land. This has unfortunately led to an important increase in abandoned buildings, which have considerable historical and architectural value.

5. Conclusions

In the last decades the area of Forenza has undergone significant land use changes. The rural landscape was subjected to significant, and often damaging, modifications in addition to changes resulting from new agricultural and environmental policies. These dynamics, in order to be correctly assessed, require detailed studies, advanced by computer technologies that can be used as decisive tools for the creation of appropriate instruments for the representation and simultaneous management of all landscape aspects. GIS allows the implementation of historical models, for quantitative and qualitative studies, in order to understand the evolution of landscape, with the aim to address appropriate spatial planning, as well as proper land use management policies. In this paper the historical analysis showed that significant changes occurred in terms of land use, during a period of about two centuries.

Compared to the situation in 1829, the natural area of Forenza municipality decreased, overall. However, during the last decades the re-vegetation of the abandoned agricultural areas occurred, a common phenomenon widespread in the south of Italy and, particularly, in the Basilicata region. Many studies show the consequences of the grassland, pastures and woodland recovery after land abandonment. The major problems related to this phenomenon have been identified: biodiversity loss, increase of fire frequency and intensity, soil erosion; and desertification, loss of cultural and/or aesthetic values, reduction of landscape diversity and of water provision [25,40]. The extraordinary properties of representation offered by GIS technologies increase the perception of the study areas and improve the informational aspects and the opportunities of visual simulation of the land use evolution.

The approach which was implemented herein may be equally applied to other areas in the Eastern Mediterranean in which there is the availability of datasets which could be analysed within a GIS, so as to assess the changes that have occurred over time in those landscapes. The areas of Eastern Mediterranean basin are indeed being profoundly transformed by human activity [40]. Agricultural lands, evergreen woodlands and maquis habitats, that dominate the present landscape of the Mediterranean basin, are the result of anthropogenic disturbances over millennia [41]. Many common factors contribute to landscape modifications: the abandonment of the countryside in the last decades is due to socio-economic changes occurred during the last decades; these changes were linked with the shift from traditional agricultural systems, characterized by a high landscape complexity, to new agricultural schemes supported by modern technology [42]. The comparison between past and present land use represents a straightforward approach for identifying landscape changes and quantifying the relationships among anthropogenic, agricultural, and forest systems, as well as fostering sound land use policies to other regions and local situations in this common territorial context.

Author Contributions: The design and conduct of this research is equally shared among the authors, who jointly performed the bibliographic analysis, the experimental study, and the discussion of the results.

Conflicts of Interest: The authors declare no conflict of interest.

References

1. Rescia, A.J.; Willaarts, A.B.; Schmitz, M.F.; Aguilera, P.A. Changes in land uses and management in two Nature Reserves in Spain: Evaluating the social-ecological resilience of cultural landscapes. *Landsc. Urban Plan.* **2010**, *98*, 26–35. [CrossRef]
2. Statuto, D.; Tortora, A.; Picuno, P. Spatial modeling and image processing of historical maps for rural landscape planning. In Proceedings of the International Conference of Agricultural Engineering-EurAgEng 2014, Zurich, Switzerland, 6–10 July 2014.
3. Pelorosso, R.; Leone, A.; Boccia, L. Land cover and land use change in the Italian central Apennines: A comparison of assessment methods. *Appl. Geogr.* **2009**, *29*, 35–48. [CrossRef]
4. Schneeberger, N.; Bürgi, M.; Kienast, P.D.F. Rates of landscape change at the northern fringe of the Swiss Alps: Historical and recent tendencies. *Landsc. Urban Plan.* **2007**, *80*, 127–136. [CrossRef]
5. Yeh, C.T.; Huang, S.L. Investigating spatiotemporal patterns of landscape diversity in response to urbanization. *Landsc. Urban Plan.* **2009**, *93*, 151–162. [CrossRef]
6. Statuto, D.; Tortora, A.; Picuno, P. Analysis of the evolution of landscape and land use in a GIS approach. In Proceedings of the First International Symposium Article on Agricultural Engineering—ISAE 2013, Session VI, Belgrade, Serbia, 4–6 October 2013; pp. 25–33.
7. Mutoko, M.C.; Hein, L.; Bartholomeus, H. Integrated analysis of land use changes and their impacts on agrarian livelihoods in the western highlands of Kenya. *Agric. Syst.* **2014**, *128*, 1–12. [CrossRef]
8. Tveit, M.; Ode, A.; Fry, G. Key visual concepts in a framework for analyzing visual landscape character. *Landsc. Res.* **2006**, *31*, 229–255. [CrossRef]

9. Franch-Pardo, I.; Cancer-Pomar, L.; Napoletano, B.M. Visibility analysis and landscape evaluation in Martin river cultural park (Aragon, Spain) integrating biophysical and visual units. *J. Maps* **2017**, *2*, 415–424. [CrossRef]
10. Domon, G. Landscape as resource: Consequences, challenges and opportunities for rural development. *Landsc. Urban Plan.* **2011**, *100*, 338–340. [CrossRef]
11. Ales, R.F.; Martin, A.; Ortega, F.; Ales, E. Recent changes in landscape structure and function in a Mediterranean region of SW Spain (1950–1984). *Landsc. Ecol.* **1992**, *7*, 3–18. [CrossRef]
12. García-Ruiz, J.M.; Lasanta, T.; Ruiz-Flano, P.; Ortigosa, L.; White, S.; Gonzales, C.; Martì, C. Land-use changes and sustainable development in mountain areas: A case study in the Spanish Pyrenees. *Landsc. Ecol.* **1996**, *11*, 267–277. [CrossRef]
13. Debussche, M.; Lepart, J.; Dervieux, A. Mediterranean landscape changes: Evidence from old postcards. *Glob. Ecol. Biogeogr.* **1999**, *8*, 3–15. [CrossRef]
14. MacDonald, D.; Crabtree, J.R.; Wiesinger, G.; Dax, T.; Stamou, N.; Fleury, P.; GutierrexLazpita, J.; Gibon, A. Agriculture abandonment in mountain areas of Europe: Environmental consequences and policy response. *J. Environ. Manag.* **2000**, *59*, 47–69. [CrossRef]
15. Santos, C.P. Succession of breeding bird communities after the abandonment of agricultural fields in southeast, Portugal. *Ardeola* **2000**, *47*, 171–181.
16. Lambin, E.F.; Geist, H.J.; Lepers, E. Dynamics of land use and land cover change in tropical regions. *Annu. Rev. Environ. Resour.* **2003**, *28*, 205–241. [CrossRef]
17. Covas, R.; Blondel, J. Biogeography and history of the Mediterranean bird fauna. *Ibis* **1998**, *140*, 395–407. [CrossRef]
18. Picuno, P.; Tortora, A.; Capobianco, R.L. Analysis of plasticulture landscapes in Southern Italy through remote sensing and solid modelling. *Landsc. Urban Plan.* **2011**, *100*, 45–56. [CrossRef]
19. Statuto, D.; Tortora, A.; Picuno, P. A GIS approach for the quantification of forest and agricultural biomass in the Basilicata region. *J. Agric. Eng.* **2013**, *XLIV*, 627–631. [CrossRef]
20. Statuto, D.; Picuno, P. Valorisation of vernacular farm buildings for the sustainable development of rural tourism in mountain areas of the Adriatic-Ionian macro-region. *J. Agric. Eng.* **2017**, *48*, 21–26. [CrossRef]
21. De Philippis, A. *Classificazione ed Indici del Clima in Rapporto Alla Vegetazione Forestale Italiana*; Società Botanica Italiana: Firenze, Italy, 1937.
22. Blakemore, M.J.; Harley, J.B. Concepts in the history of cartography. A review and perspective. *Cartographica* **1980**, *17*, 1–120.
23. Laxton, P. The Geodetic and topographical evaluation of English county maps, 1740–1840. *Cartogr. J.* **1976**, *13*, 37–54. [CrossRef]
24. Jenny, B.; Hurni, L. Studying cartographic heritage: Analysis and visualization of geometric distortions. *Comput. Graph.* **2011**, *35*, 402–411. [CrossRef]
25. Tortora, A.; Statuto, D.; Picuno, P. Rural landscape planning through spatial modelling and image processing of historical maps. *Land Use Policy* **2015**, *42*, 71–82. [CrossRef]
26. Yang, Y.; Zhang, S.; Yang, J. A review of historical reconstruction methods of land use/land cover. *J. Geogr. Sci.* **2014**, *24*, 746–766. [CrossRef]
27. De Marco, D. *La Statistica Del Regno Di Napoli Nel 1811—Volume III (1900)*; Accademia Naz. dei Lincei: Rome, Italy, 1990; pp. 24–25, 54–59, 228–233, 474–477.
28. Forenza Municipality. *Relazione Sulle Usurpazioni Del Bosco Demaniale Di San Giuliano*; Stab. Tipografico "Fulgur": Potenza, Italy, 1915.
29. Lafreniere, D.; Rivet, D. Rescaling the Past through Mosaic Historical Cartography. *J. Maps* **2010**, *6*, 417–422. [CrossRef]
30. Boùùaert, M.C.; De Baets, B.; Vervust, S.; Neutens, T.; De Maeyer, P.; Van de Weghe, N. Computation and visualisation of the accuracy of old maps using differential distortion analysis. *Int. J. Geogr. Inf. Sci.* **2016**, *30*, 1225–1280.
31. Statuto, D.; Cillis, G.; Picuno, P. Analysis of the effects of agricultural land use change on rural environment and landscape through historical cartography and GIS tools. *J. Agric. Eng.* **2016**, *47*, 28. [CrossRef]
32. Lautenbach, S.; Kugel, C.; Lausch, A.; Seppelt, R. Analysis of historic changes in regional ecosystem service provisioning using land use data. *Ecol. Indic.* **2011**, *11*, 676–687. [CrossRef]

33. Kosztra, B.; Arnold, S. *Proposal for Enhancement of CLC Nomenclature Guidelines*; ETC/SIA Deliverable EEA Subvention 2013 WAI Task 261_1_1: Applying EAGLE Concept to CLC Guidelines Enhancement. European Environment Agency Internal Report; EEA-European Environment Agency: Copenhagen, Denmark, 2013.
34. Kakiuchi, T.; Chikatsu, H. Reconstruction and visualization of original landscape on reverse mapping. *Int. Arch. Photogram. Remote Sens. Spat. Inf. Sci.* **2010**, *38*, 311–315.
35. Picuno, P. Vernacular farm buildings in landscape planning: A typological analysis in a southern Italian region. *J. Agric. Eng.* **2012**, *43*, 130–137. [CrossRef]
36. Italian Regulations for the Expropriation, Land Reclamation, Processing and Allocation of the Peasants's Land. In *G.U. No. 249, 28/10/1950*; Law of October 21, No. 841; 1950. Available online: http://www.edizionieuropee.it/LAW/HTML/10/zn2_02_033.html (accessed on 18 September 2017).
37. Commission Regulation EEC No. 1272/88 Laying Down Detailed Rules for Applying the Set-Aside Incentive Scheme for Arable Land. Available online: http://research.omicsgroup.org/index.php/Set-aside (accessed on 18 September 2017).
38. Van Vliet, J.; de Groot, H.L.F.; Rietveld, P.; Verburg, P.H. Manifestations and underlying drivers of agricultural land use change in Europe. *Landsc. Urban Plan.* **2015**, *133*, 24–36. [CrossRef]
39. Gutzler, C.; Helming, K.; Balla, D.; Dannowski, R.; Deumlich, D.; Glemnitz, M.; Knierim, A.; Mirschel, W.; Nendel, C.; Paul, C.; et al. Agricultural land use changes—A scenario-based sustainability impact assessment for Brandenburg, Germany. *Ecol. Indic.* **2015**, *48*, 505–517. [CrossRef]
40. Geri, F.; Amici, V.; Rocchini, D. Human activity impact on the heterogeneity of a Mediterranean landscape. *Appl. Geogr.* **2010**, *30*, 370–379. [CrossRef]
41. Blondel, J. The 'design' of Mediterranean landscapes: A millennial story of humans and ecological systems during the historic period. *Hum. Ecol.* **2006**, *34*, 713–729. [CrossRef]
42. Rühl, J.; Pasta, S.; La Mantia, T. Metodologia per lo studio delle successioni secondarie in ex-coltivi terrazzati: Il Caso studio di Pantelleria (Canale di Sicilia). *For. J. Silvic. For. Ecol.* **2005**, *2*, 388–398. [CrossRef]

land

MDPI

Review

Transferring Landscape Character Assessment from the UK to the Eastern Mediterranean: Challenges and Perspectives

Geoffrey Griffiths

Department of Geography & Environmental Science, The University of Reading, Reading RG6 6AB, UK; g.h.griffiths@reading.ac.uk; Tel.: +44-118-378-8737

Received: 21 January 2018; Accepted: 5 March 2018; Published: 15 March 2018

Abstract: Landscape character assessment (LCA) has a significant contribution to make as a spatial framework for the emerging concept of 'multi-functional landscapes', a landscape providing a range of functions, services, and human-derived benefits. The paper reviews the development of LCA in Northwest Europe with a brief description of more recent LCA projects in a Mediterranean context. This is followed by a comparative description of the Living Landscapes approach developed in the UK as applied to Cyprus. The focus is upon the challenges, and limitations, of transferring a method developed in one context to the different physical and cultural setting of the island of Cyprus examining differences in the definition of landscapes, the availability of information on the cultural landscape, the importance of incorporating a strong element of 'time-depth', and the potential of LCA for enhancing land use policy at a time of increased land pressures in the Mediterranean.

Keywords: landscape character assessment; multi-functional landscapes; planning; UK; Cyprus

1. Introduction

'Landscape' means different things to different people in different contexts and, in some parts of the world, is not even recognized as a concept. However, increasingly we recognize some idea of 'landscape' as a convenient framework within which to plan and manage land resources. For example, the concept of a multi-functional landscape has been current for some time [1,2], with the potential to deliver a range of human benefits at a time of continuing pressure on the rural landscapes of Europe. The concept has been given further credibility and application with the allied concept of ecosystem goods and services [3,4]. However, not only is the concept and definition of landscape contested [5], but the scale at which different ecosystem goods and services function, the many trade-offs between these functions, and the impact of change on the stocks and flows of natural capital [6] are all critical challenges to the application of the ecosystem services paradigm for sustainability. This challenge is made more difficult against a background of very large variability in the landscapes of Europe and, in this context, the Eastern Mediterranean, the product of settlement and cultural patterns over millennia set against a highly variable physical background of climate, soils, geology, and topography. Whilst this variability is acknowledged with many attempts to capture it using a range of mapping methods at a variety of scales [7,8], to date there is only limited literature, including in the geographical literature, that recognises the importance of 'landscape' in the measurement and provision of ecosystem goods and services for human well-being [9] and for physical planning. Thus, Landscape character assessment (LCA) has a potentially key role in this respect bridging the gap between people and landscape, on the one hand, and planning and the need to protect and enhance natural capital on the other. This paper reviews, briefly, the background of LCA in the UK and its recent application in a completely different environment, the island of Cyprus in the Eastern Mediterranean. The principal aim of the paper, however, is to evaluate the extent to which a system developed in one part of Europe within the highly-institutionalised framework of the UK (in this example), can be transferred and applied not only to a different physical environment, but also to a country with a very different system of planning and governance. The paper starts with a general introduction to LCA in a pan-European context, followed by more specific comments on the development of LCA in the UK before a description of the application of LCA in Cyprus. Concluding sections focus on the challenges of transferring a method developed in the UK to the different environment of the Eastern Mediterranean.

2. Landscape character assessment (LCA)

2.1. Background

Without the natural and cultural variability of landscapes being incorporated into a coherent spatial framework, it is unlikely that the potential of the ecosystem services (ES) approach will ever be realised [4]. Effective physical planning for development control and environmental assessment (EA) for example, depends upon the availability of a framework of spatial units that captures differences in 'place'. The system needs to be applicable at a range of scales, transferable between places (countries/regions) and sufficiently practical and understandable for ease of application rather than simply an academic exercise in the recording and mapping of landscapes. This poses very specific challenges and questions: is a universal, widely-applicable system necessary at the European scale or do we continue with the present ad hoc system of national, even sub-national approaches? How is the spatial framework to be constructed: at one end of the debate is the view famously expressed by the European Landscape Convention [10]) that a landscape is, 'an area, perceived by people, whose character is the result of the action and interaction of natural and/or human factors' to apparently objective computer-generated maps based on statistical analysis of spatial input data [7,8].

It is argued that the advantage of adopting an integrated landscape approach is that it incorporates the entire range of meanings that we attribute to landscapes as a "spatial entity; a mental entity; a temporal dimension and as a nexus of nature and culture" [11]. The danger is that the wide meanings attached to landscape prohibit its practical application as a spatial framework at a time when there is an urgent need to plan for resilient landscapes under continuing pressure from threats, such as, climate change, rapid infrastructure development, and changes in agriculture, including land abandonment, particularly prevalent in Mediterranean countries. In the UK, at least, planning has moved from development control towards spatial planning with a new emphasis on sustainability as a core concern [5]. Most counties (and districts) in England have commissioned a Landscape character assessment, many of which are being adopted for use in planning [12]. However, too many of these landscape assessments are visually based, focusing on the visible, natural attributes of the landscape, and often neglecting the cultural aspects that also contribute significantly towards landscape distinctiveness. Without informed map analysis, the assessment simply reflects the visual dimension of the landscape losing the all-important understanding of the contribution of natural factors and cultural processes that help to explain contemporary patterns. Such a visual focus aligns closely with a place-based approach, but stands in sharp contrast to the spatial mapping of landscape units with a strong focus on ecological and environmental objectives that continue to be developed by geographers, landscape ecologists, and historical ecologists. This tension remains real and problematic in the sense that, in some contexts, principally at local and regional scales, it is constraining the development and application of methods for more sustainable planning. Nevertheless, both types of approach are necessary if the system is to be both useable, in the sense of boundaries on a map, and perceptual, in the sense of 'place-based' and recognisable and appreciated by people. Thus, the drawing of borders and the categorisation of areas, e.g., landscape types, are the mapped spatial units that "aim to steer, control and guide how the landscape develops, and is central in understanding what constitutes integration" [13]. This, of course, comes with its own attendant dangers of quasi-objectivity: the perception that a line on a map is 'truth' rather than a spatial simplification or generalization of the diversity of 'land'. By contrast, landscape is often only viewed as 'real' if it is perceived through the lens of perception and association that makes different landscapes recognisable and distinctive depending upon association and understanding. As an example of this tension, policies to protect and enhance biodiversity need to be sensitive to the historical forces that have shaped distinctiveness, often overlooked, for example, in current debates about rewilding or reforestation [14].

This failure to agree and develop a common and consistent approach is, therefore, the consequence of many inter-related factors. There is a critical distinction between 'environmental stratification' and LCA that continues to confuse researchers and practitioners alike. The availability of mapped information in digital form, increasingly from the Internet, enables users to adopt statistical, parametric techniques to develop a landscape typology whose properties are critically determined by the user selection of input variables and the processing algorithm [15]. This is useful when the goal is a classification of the whole of Europe, providing semi-automated techniques that rely upon known controls of landscape type (elevation, soil and geology type, land cover) to classify over very large areas. Other good examples include the UK Land Classification System, one of the earliest attempts at environmental stratification as a spatial framework for monitoring countryside change [16]. Many modelling applications benefit from this 'gridded' approach that links well with information derived from satellite imagery and provides a uniform unit area in which to calculate key variables for model input. However, the grid as a spatial framework often fails to capture the 'grain' of the landscape imposing a rigid system onto a resource that is continuously variable, both physically and culturally. Perhaps equally important, people do not resonate with gridded cells—they are not landscapes and, therefore, are meaningless in the context of any sense of identity or sense of place.

By contrast, LCA needs to be more holistic, starting with the construction of a spatial framework that is gradually 'filled' as more detailed information becomes available [17]). The emphasis is on the integration of both natural and cultural information and relies, although not exclusively, on informed understanding of historical and environmental controls on landscape evolution as part of the visual interpretation and integration of mapped sources of data. In this way landscape assessment captures the 'scale' of the landscape, identifying regions of relatively uniform physical and cultural character, often of substantially different geographical extent. There is also the danger of 'repeatability', different interpreters coming to different conclusions to produce a typology and a hierarchy of spatial units that may not be comparable. Nevertheless, the process of mapping is instructive, enabling the interpreter to develop a better understanding of the evolution and inherent character of a landscape.

The availability of data critically determines the type of method adopted. In parts of Europe, where there is a strong tradition of enquiry into historical geography/landscape history and where mapped historical information is often widely available, this provides a strong basis for inclusion of cultural information in LCA.

Finally, there is still confusion about the meaning of some of the terms that are used to describe the landscape characterisation and decision-making process. The temptation to present and reproduce maps without a clear explanation of the methods of classification, the number, scale, and currency of the input variables and the original purpose of the mapping, all contribute towards a generally poor understanding of the process and purpose of LCA. This reduces the value of LCA as a tool for both planning and land management decisions and constrains the intellectual quality of the debate about future development at a time when this is urgently needed.

2.2. LCA in the Mediterranean

It is not the case that there has been no interest in the potential of LCA in a Mediterranean context, although a detailed review of what has been achieved and where is beyond the remit of the present paper. Vogiatzakis [18] in particular, has provided a useful summary citing early examples from Spain, Italy (specifically Sardinia), Portugal, France, Slovenia, Malta, and Cyprus. He makes the point that cultural factors have been as significant as natural ones in the development of Mediterranean landscapes [19]. Equally important, Vogiatzakis [18] argues that it is the mapping of the cultural component, particularly in relation to farming systems and associated field patterns, which is missing due to the absence of relevant datasets and a tradition of mapping historical landscapes. More recently, a review of LCA in the Eastern Mediterranean [20] identified and mapped a total of 69 landscape types, of which 18 were classified as rare resulting from specific geomorphology or intensive anthropogenic activities. The significance of this work is that it is the first step towards a standard typology applicable across a large part of the Mediterranean providing consistency between, and within, countries and, potentially, facilitating trans-boundary cooperation for the long-term protection and enhancement of important cultural landscapes and areas of high nature conservation importance.

2.3. The Landscape Map of Cyprus

The obvious differences in the physical geography of Cyprus, its cultural history, and the contemporary pressures it faces from development, provide a useful contrast to the UK where LCA has been under development and application for more than 20 years. The aim of the Cyprus project was to produce a Landscape character assessment of the whole island based on the Living Landscapes Project (LLP) approach [12]) for spatial planning. There was, therefore, a focus on a 'top-down' mapping approach based on the integration of layers of map information to generate discrete landscape units. This had to be defensible legally in a planning context and, therefore, more likely to be adopted at a time of rapid land-use change and continuing development pressure. These remain urgent issues in Cyprus and it is anticipated that the landscape framework will be used to identify locations that are highly sensitive to change, for example, those areas that have strong ecological, cultural, or visual distinctiveness. In the case of Cyprus, the existing spatial planning system provides the required efficiency, transparency, and citizen participation for the successful integration of landscape considerations into spatial planning, but this needed to be supported by a landscape framework to be an effective decision-making tool. Only with the incorporation of landscape character/quality objectives into development plans can real progress be made in implementing the ELC [10]. However, before spatial plans could be populated with landscape character/quality objectives, and before the planning system could refer to effective landscape policies, a crucial step had to be taken towards the place-based assessment of landscape character and diversity throughout the island. The output of this exercise within the context of the EU-funded MedScapes project [21] was to produce a provisional map of landscape character types (LCTs) for Cyprus at a broad (Level 1) and, later, at the more detailed Level 2 scale. The Level 2 map was subsequently updated with additional information on the cultural components of the landscape from map sources and fieldwork, as part of validation and evaluation.

The hierarchical nature of the Living Landscapes approach [12] enabled the broad patterns of the natural physiography of the island to be mapped based on four attributes: physiography (geology structure and landform); ground type (rock type and soil type); landform (topography from contours); and land cover (Table 1).

Table 1. Living Landscape Project, definitive attributes of landscape description units at Level 1 and Level 2 [11].

	Regional Level 1 (1:250,000)	Sub-regional Level 2 (1:50,000)
Natural	Physiography	Landform Geology
	Ground type	Geology (rock type) Soils
Cultural	Land cover	Tree cover Farm type (cover)
	Settlement	Settlement Farm type (structure)

A series of map overlays at approximately 1:250,000 scale were produced for the whole island derived from digital maps provided by the Cypriot authorities and stored within a GIS (geographical information system). Spatial data-sets were obtained from a variety of sources including the Department of Town Planning in Cyprus:

- CORINE Land Cover (http://reports.eea.europa.eu/COR0-landcover/en);
- Cyprus Soil Map (1:250,000 scale);
- Cyprus Geology Map (1:250,000 scale) (Cyprus Geological Survey);
- Cyprus 20 m contours (Cyprus Department of Lands and Surveys); and
- Cyprus Topographic Maps (1:50,000) (Cyprus Department of Lands and Surveys).

By overlaying elevation data (contours) on top of the geological/soils data the island was divided initially into broad physiographic units. These units were then overlaid on the ground type/land cover data and subdivided, where appropriate, to distinguish the ecological character and inherent capability of the land. The spatial units that emerge provide the framework for understanding and mapping the historic pattern of land use and settlement, where this information is available. A total of 21 LCTs were mapped at Level 1 (Figure 1 and Table 2), reflecting the specific natural and cultural aspects of the Cypriot landscape.

From Table 2 it is immediately clear that the derivation of landscape character types at the broad Level 1 scale, is linked to the specific conditions of land cover, settlement and geology/soil types characteristic of Cyprus. Information about historical patterns (e.g., land use, settlement patterns, field size, and shape) are of critical significance for mapping the cultural dimension of landscape character but, unlike the situation in a UK context, were not available for Cyprus and had to be generated separately. However, the availability of 1:50,000 scale topographic maps from 1975 (i.e., before the rapid urbanisation of recent decades) provided the opportunity to interpret and map differences in settlement and field patterns for incorporation into the typology at the next, more detailed level in the hierarchy (Level 2). More than 200 spatial units were mapped at Level 2 (Table 2). This proved to be one of the most challenging aspects of the project and, based on extensive fieldwork, a scheme was developed enabling differences in field patterns to be mapped based on the specific conditions prevalent in Cyprus. This represented a critical variation from the typology developed in a UK context [12], but it is known that differences in field patterns are diagnostic of the often complex interrelationship between historical and natural factors operating across the island.

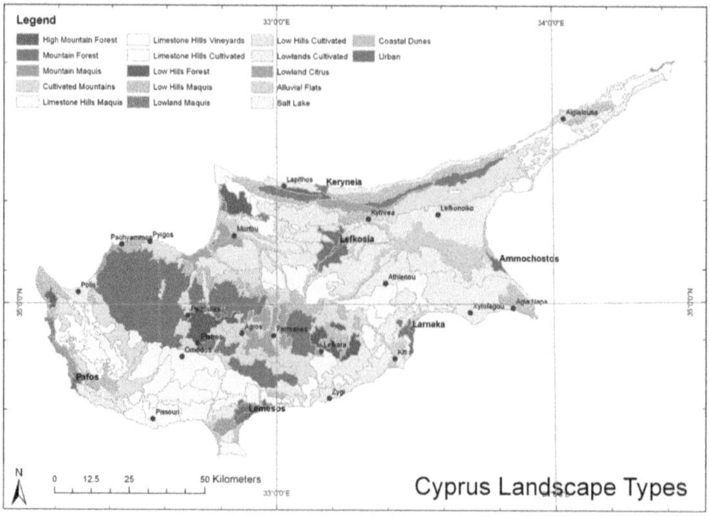

Figure 1. The Level 1 landscape character types for Cyprus.

Table 2. Landscape character types (LCTs) at Level 1 within the number of Level 2 units in each LCT.

LCT	Number of Level 2 LDUs	Description
Coastal dunes	1	Low hills of unconsolidated sand, associated with a strip of bare sand/pebbles, along the coast; unsettled, 'wild' landscape with a covering of shrubs and other semi-natural vegetation, although patches of bare ground are a feature in places
Coastal Wetlands	3	Mostly dominated by salt lakes: Landscape with bare land in an undrained natural basin, periodically inundated with salt water and associated ecosystems, like salt marshes
Settled coastal hill shrublands	4	Undulating hilly topography with covering of shrubs and other semi-natural vegetation; settled landscape lying along the coastline
Settled coastal lowland shrublands	4	A low-lying landscape near the coast dominated by Mediterranean shrubland; settled
Settled cultivated alluvial plains	11	Settled, cultivated lowland landscape, mostly low lying, with a gently rolling topography
Settled cultivated coastal alluvial plains	7	A flat, settled low-lying landscape, associated with seasonal river channels and their flat plains. Much of the land is cultivated, but patches of pastoral grazing land are a feature in places
Settled cultivated coastal hills	2	An undulating, settled and cultivated landscape of low (below 500 m) hills, with patches of shrubby semi-natural vegetation lying along the coastline
Settled cultivated coastal lowlands	7	Settled and cultivated lowland landscape near the coast, with discrete rural villages
Settled cultivated hills	20	An undulating, settled and cultivated landscape of low (below 500 m) hills, with patches of shrubby semi-natural vegetation
Settled cultivated lowlands	28	A settled and cultivated lowland landscape, with discrete rural villages. Although mostly low lying, with a gently rolling topography, this landscape also includes a scattering of prominent, small hills
Settled cultivated mountains	5	A settled and cultivated landscape on higher ground, with an undulating, in places steeply sloping topography, often covered with narrow, stone walled terraces with mixed cultivation, including a greater or lesser proportion of fruit orchards and vineyards, which thrive on the mineral rich volcanic soils. Small towns and villages are a characteristic feature of this landscape, often situated along steep valley sides. Houses are stone built with tiled roofs
Settled cultivated valleys	8	Valleys with relatively medium slopes, characterised by agricultural uses; largely settled
Settled forested hills	2	A heavily wooded, settled, highland landscape associated with steeply, sloping, high (mostly above 500 m) hills
Settled forested mountains	4	A heavily wooded, settled, highland landscape associated with steeply/sloping, high (mostly above 500 m) mountains, found typically on resistant volcanic (e.g., Troodos) or metamorphic (e.g., Pentadactylos) geology
Settled forested valleys	2	Valleys with natural forest vegetation on the sloping sides of a river with permanent flow; usually settled
Settled hill shrublands	18	Undulating hilly topography with covering of shrubs and other semi-natural vegetation; settled landscape
Settled hill with vineyards	5	An undulating, in places steeply sloping, cultivated limestone landscape, with discrete rural villages and frequent vineyards, often located along narrow, stone walled terraces
Settled mountain shrublands	5	Steeply sloping, settled, high (mostly above 500 m) mountains with a covering of shrubs, typically, the endemic Quercus alnifolia above 100 m, and other semi-natural vegetation. Prominent rocky outcrops and cliffs are a feature of this landscape in places
Settled rangeland hills	2	A heavily grazed landscape associated with valley hillsides; settled with limited coverage of natural vegetation
Unsettled coastal hill shrublands	4	Undulating hilly topography with covering of shrubs and other semi-natural vegetation; strongly unsettled landscape lying along the coastline
Unsettled coastal lowland shrublands	4	A low-lying landscape near the coast dominated by Mediterranean shrubland; unsettled

3. Mapping and Interpretation

The apparent dichotomy between landscape as diffuse 'scenic space' and landscape as sharply defined, apparently objective lines on a map does not suggest that an adherence to the ELC [10] approach is a pre-requisite for LCA, simply whether a system that relies entirely on the generation of mapped spatial units is also able to accommodate the place-based appreciation of landscape perceived by people as scenic, cultural space. This remains an open question and one of continuing academic debate, but, in a Cypriot context, the rapid rate of development, and the critical need to protect and enhance scenic beauty and ecosystem services, presupposes the existence of a framework of spatial units derived from a commonly understood process, using data from known sources with some degree of quality assurance legally defensible in a planning context.

Data, therefore, becomes a critical issue: its availability in digital format; modernity; type; and scale. Increasingly, the availability of physical data (soils geology, climate; land cover, elevation, etc.) is becoming available from Internet sources. A good example is the frequently-updated land cover mapping available via the CORINE project from the European Environment Agency (EEA). In relation to mapped cultural information, however, the situation is very different. The UK is fortunate in being well-mapped and recorded with, for example, national maps of historical settlement patterns [22]. In many parts of Europe where there is a strong tradition of enquiry into historical geography/landscape history, and where mapped historical information is often widely available, this provides a strong basis for inclusion of cultural information in LCA. Recent work in the UK on historic landscape characterisation [23], and a long tradition of enquiry in historical ecology and landscape history [24–26], has contributed significantly towards the incorporation of cultural information into the landscape character assessment process. This is not a situation that is typical of other parts of Europe and, certainly in an Eastern Mediterranean context, this proved to be the case: much of this information had to be generated from primary map sources to generate proxy information on cultural attributes, e.g., field and settlement patterns. This also raises the important question of transferability: patterns in the landscape that are diagnostic of processes of historical evolution in one context, or country/region, may not be transferrable to another.

Time-Depth

Peoples' perception, understanding, and appreciation of place and its distinctiveness is critically determined by 'time-depth', the visual expression of the cultural forces that have shaped a landscape over time. This is often visually observed, filtered through a complex set of individual experiences and understanding. However, in no other aspect of landscape assessment does the inherent contradiction between subjective appreciation of landscape and the need for 'objective' mapping collide quite so strongly as in the definition and analysis of 'time-depth'. As a result, there is the danger that our understanding of the processes that shape the strong visual patterns in one landscape when transferred to a different historical context, will be misplaced. In a UK context, for example, the distinctive pattern of small, irregular fields so characteristic of the 'ancient countryside' does not translate to Cyprus where small, irregular fields are determined by a separate set of historical processes and topographical influences. This may not matter in a visual sense: the two landscape types show a similarity of pattern. It matters strongly in our understanding and appreciation of landscape and the way in which this affects our ability to measure landscape sensitivity and, ultimately, its capacity to absorb change.

In this context in England, the essential 'time-depth' component of landscape is provided by historic landscape characterisation (HLC), a mapping approach that captures the evolution of a landscape based on archival mapping in a digital GIS environment. The results from HLC are being used to update LCA mapping in England by using the HLC data to identify key historical events, for example, parliamentary enclosure of the medieval open-field system, to validate the cultural attributes of the LCA maps. Typically, the HLC maps express the land cover of a region linked to the dominant historical process that 'produced' that landscape, for example 'anciently-enclosed land', 'industrial land', and 'recently-enclosed land' [27]. The question arises, therefore, whether a similar approach could be developed in Cyprus, and the Mediterranean in general, based on existing maps and information. The answer, of course, depends upon the availability of archived information on historical land changes and the expertise to interpret and map them. In Cyprus there is a wealth of information recorded and mapped by successive occupiers, including the Ottomans and British, the cultural layers relating to land ownership, settlement, systems of taxation, etc., that have shaped the landscapes of contemporary Cyprus now changing rapidly in response to the demands of mass tourism.

4. Planning for Change: Landscape Sensitivity and Capacity

The contemporary landscape of Cyprus is being shaped by the new 'occupation': mass tourism, contributing to more than 12% of GDP demanding hotels, communications, and other development associated with membership of the EU and an unusually favourable climate, even by Mediterranean standards. This continues to exert a profound impact on Cypriot landscapes visually, ecologically, and culturally. Whilst the MedScapes project [21] was a first attempt to comply with the European Landscape Convention [10], with a remit focused strictly on mapping, it was clear at the time of the project that the obvious next step would be to apply the framework for evaluation of the sensitivity and capacity of different landscape types.

Against this background of rapid urbanisation the protection of important landscapes depends on the development of a coherent and defensible landscape policy supported by our understanding of which landscapes are highly sensitive to what types of change. However, the use and understanding of landscape sensitivity and capacity remains vague reflecting the context-dependent nature of both terms. Natural England's paper, *Techniques and Criteria for Judging Capacity and Sensitivity* [28], provides the best guidance however, on the capacity of landscapes to absorb change in relation to measures of their relative sensitivity. Critically, landscape sensitivity is referred to at two different scales: firstly, overall sensitivity, relating to the inherent character of the landscape in question, and, secondly, the specific sensitivity, to a defined proposal (type) of change.

An important objective of the Cyprus landscape project was to develop and apply measures of landscape sensitivity and capacity as part of the development of landscape policy. To this end the more detailed Level 2 mapping based on eight definitive attributes (Table 1) represents a sub-division of the island that captures subtle differences in character in which to determine the ecological, cultural, and visual sensitivity of each landscape unit differently and to evaluate its capacity for absorbing change. This was only possible with extensive field observations of the 'condition' of each landscape unit that, itself, depends on a coherent set of spatial units that can be visited and sampled in a systematic way. Field survey was undertaken in Cyprus to visit every Level 2 LDU to assess their condition in 2015. The data are currently being used in the Akamas peninsula in Cyprus as part of the development of land use policy for this region of high biodiversity and natural beauty given the significant pressures for new development.

Whilst physical planning dominates land use policy in Cyprus currently, this is an island where other pressures are also impacting on the functioning of ecosystems and the long-term integrity of landscapes. Forecasts from regional climate model simulations for the 21st century, for example, highlight the vulnerability of Cyprus to climate change by projecting an increase in maximum temperature of 1.3–1.9 °C for 2021–2050 and 3.6–5 °C for 2071–2100, and a sharp decrease in rainfall by the end of the century [29]. Rural to urban drift is resulting in extensive agricultural land abandonment in remote areas, in common with many other parts of the Mediterranean. These, and other, impacts are changing the nature of landscapes in ways that are equal in magnitude to any process in the historical past and, as such, are impacting directly on the ecosystem services provided by such landscapes. The concept of ecosystems services [30] has become an important model for linking the functioning of ecosystems to human welfare [31]. Such a link presupposes a spatial framework within which to map and evaluate the provision of ecosystem services at scales ranging from the local to the global [32,33]. In this context a critical requirement is that landscape units, often referred to as 'service providing units' (SPUs) [34] are characterised by defined, recognisable units. Some have argued that we need a system that incorporates units that both generate ecosystem services (service providing areas, SPAs) and units that receive those benefits, so-called service benefit areas (SBAs) [35].

However, spatial correlations between ecosystem services and landscape types have often been assumed rather than demonstrated, with the possible notable exception of the paper by Chan et al. [36]. There is a need to develop case studies across a range of landscape types, such as the Cyprus landscape study, to demonstrate the value of LCA as a spatial framework for capturing ecosystem services. This type of framework of functional units also offers the potential to develop useful metrics of the habitats, landscape elements, and land use constraints that significantly influence the generation of ecosystem services and the benefits they offer humanity [35].

5. Discussion and Conclusions

Despite the great cultural and natural diversity of Mediterranean landscapes, they are faced with common threats, including climate change, pollution, overgrazing, and tourism development. These problems, combined with a lack of public awareness and political commitment, are hindering the protection and sustainable planning of natural and cultural landscapes. In terms of land use, this absence of 'landscape' from national and regional policy, and the lack of a clear vision for landscape planning, may be a strong disincentive towards the development and implementation of future landscape mapping projects. Thus, future progress will be dependent upon the development and validation of a system of landscape mapping; that is:

- characterised by a strong spatial dimension of mappable units;
- incorporates both natural (soils, geology landform) and cultural dimensions;
- hierarchical, with different nested levels in the system providing richer content at large scales over smaller areas, and vice versa;
- a consistent and robust typology at all scales;
- contains a strong 'time-depth' dimension, capturing the evolution of a landscape;
- 'fits' the existing institutional and political planning framework; and
- transferability between scales and across space, based on consistent data and methods.

The experience from Cyprus indicates that the LCA approach developed in the UK [12] could be successfully transferred to other contexts with a different cultural and natural environment. However, whilst the method proved to be transferrable, the map layers did not, reflecting the absence of mapping of sufficient detail and quality to capture the major historical, 'moments' in the evolution of Cypriot landscapes, and the markedly different physical background and climate of Cyprus compared to the UK.

Nevertheless, despite the LCA approach being available for more than 20 years in Northwest Europe, it remains generally true that land use planning, ecosystem assessment, and environmental assessment (EA) tend not to depend upon a landscape framework. Instead, traditional approaches adopt the water catchment, local authority region, coastal zones, and forest zones as the principal planning unit simply reflecting the reality of existing political and administrative frameworks. Moreover, new approaches and typologies are continually being developed and tested depending upon the purpose of the application: habitat restoration, ecosystem assessment, agri-environment targeting, land-use planning, etc. The result tends to be a sectoral approach that fails to take into account the critical trade-offs between conflicting land use at a time of increasing pressures. The landscape approach, by contrast, provides a robust and consistent framework that is both 'place-based' and applicable across a range of scales and contexts.

Acknowledgments: This work was funded by the EU ENVI MedScapes project, with contributions from the University of Reading, UK. Thanks also to the Laono Foundation, Cyprus, Ioannis Vogiatzakis (Open University, Cyprus), and Phaedon Enotiades, Spatial Planning and Countryside Policy, Town Planning Dept., Cyprus.

Conflicts of Interest: The author declares no conflict of interest.

References

1. Naveh, Z. Ten major premises for a holistic conception of multifunctional landscapes. *Landsc. Urban Plan.* **2001**, *57*, 269–284. [CrossRef]
2. Fry, G.L.A. Multifunctional landscapes—towards transdisciplinary research. *Landsc. Urban Plan.* **2001**, *57*, 159–168. [CrossRef]
3. Millennium Ecosystem Assessment. *Ecosystems and Human Well-Being: Synthesis*; Island Press: Washington, DC, USA, 2005.
4. De Groot, R.S.; Alkemade, R.; Braat, L.; Hein, L.; Willemen, L. Challenges in integrating the concept of ecosystem services and values in landscape planning, management and decision making. *Ecol. Complex.* **2010**, *7*, 260–272. [CrossRef]
5. Selman, P. *Planning at the Landscape Scale*; Routledge: Abingdon, UK, 2006.
6. Costanza, R.; Daly, H.E. Natural capital and sustainable development. *Conserv. Biol.* **1992**, *6*, 37–46. [CrossRef]
7. Wascher, D.M. (Ed.) *European Landscape Character Areas—Typologies, Cartography and Indicators for the Assessment of Sustainable Landscapes*; Final Project Report as Deliverable from the EU's Accompanying Measure Project European Landscape Character Assessment Initiative (ELCAI), funded under the 5th Framework Programme on Energy, Environment and Sustainable Development (4.2.2); Wageningen University: Wageningen, The Netherlands, 2005; 150p.
8. Mücher, C.A.; Klijn, J.A.; Wascher, D.M.; Schaminée, J.H.J. A new European Landscape Classification (LANMAP): A transparent, flexible and user-oriented methodology to distinguish landscapes. *Ecol. Indic.* **2010**, *10*, 87–103. [CrossRef]
9. Potschin, M.B.; Haines-Young, R.H. Ecosystem services. *Prog. Phys. Geogr.* **2011**, *35*, 575–594. [CrossRef]
10. Council of Europe. *European Landscape Convention, Florence, European Treaty Series—No. 176*; Council of Europe: Strasbourg, France, 2000.
11. Tress, B.; Tress, G. Capitalising on multiplicity: A transdisciplinary systems approach to landscape research. *Landsc. Urban Plan.* **2001**, *57*, 143–157. [CrossRef]
12. Warnock, S.; Griffiths, G. Landscape characterisation: The living landscapes approach in the UK. *Landsc. Res.* **2015**, *40*, 261–278. [CrossRef]
13. Stenseke, M. Integrated landscape management and the complicating issue of temporality. *Landsc. Res.* **2016**, *41*, 199–211. [CrossRef]
14. Olwig, K.R. Virtual enclosure, ecosystem services, landscape's character and the 'rewilding' of the commons: The 'Lake District' case. *Landsc. Res.* **2016**, *41*, 253–264. [CrossRef]
15. Metzger, M.J.; Bunce, R.G.H.; Jongman, R.H.; Mücher, C.; Watkins, J.W. A climatic stratification of the environment of Europe. *Glob. Ecol. Biogeogr.* **2005**, *14*, 549–563. [CrossRef]
16. Bunce, R.G.H.; Barr, C.J.; Clarke, R.T.; Howard, D.C.; Lane, A.M.J. Land Classification for Strategic Ecological Survey. *J. Environ. Manag.* **1996**, *47*, 37–60. [CrossRef]
17. Van Eetvelde, V.; Antrop, M. Indicators for assessing changing landscape character of cultural landscapes in Flanders (Belgium). *Land Use Policy* **2009**, *26*, 901–910. [CrossRef]
18. Vogiatzakis, I.N. Mediterranean experience and practice in Landscape Character Assessment. *Ecol. Mediter.* **2011**, *37*, 17–31.
19. Rackham, O.; Moody, J. *The Making of the Cretan Landscape*; University Press Manchester: Manchester, UK, 1996.
20. Vogiatzakis, N.I.; Manolaki, P. Investigating the Diversity and Variability of Eastern Mediterranean Landscapes. *Land* **2017**, *6*, 71. [CrossRef]
21. Vogiatzakis, I.N.; Manolaki, P.; Trigkas, V. *LCA Training and Implementation*; Medscapes WP5 Final Report; Open University Cyprus: Lefkosia, Cyprus, 2015.
22. Roberts, B.K.; Wrathmell, S. *An Atlas of Rural Settlement in England*; English Heritage: London, UK, 2000.
23. Fairclough, G.; Sarlov-Herlin, I.; Swanwick, C. (Eds.) *Routledge Handbook of Landscape Character Assessment*; Routledge: Abingdon, UK, forthcoming.
24. Peterken, G. *Natural Woodland, Ecology & Conservation in Norther Temperate Regions*; Cambridge University Press (CUP): Cambridge, UK, 1996.
25. Hoskins, W.G. *The Making of the English Landscape*; Hodder & Stoughton: London, UK, 1969.
26. Darby, H.C. *A New Historical Geography of England*; CUP: Cambridge, UK, 1973.

27. Turner, S. Historic Landscape Characterisation: A landscape archaeology for research, management and planning. *Landsc. Res.* **2006**, *31*, 385–398. [CrossRef]
28. Swanwick, C. *Topic Paper 6: Techniques and Criteria for Judging Capacity and Sensitivity*; The Countryside Agency and Scottish Natural Heritage: Carys Swanwick, UK, 2004.
29. IPCC. *Climate Change 2014: Synthesis Report. Contribution of Working Groups I, II and III to the Fifth Assessment Report of theIntergovernmental Panel on Climate Change*; IPCC: Geneva, Switzerland, 2014.
30. Daily, G.C. *Nature's Services: Societal Dependence on Natural Ecosystems*; Island Press: Washington, DC, USA, 1997.
31. Fisher, B.; Turner, R.K.; Morling, P. Defining and classifying ecosystem services for decision making. *Ecol. Econ.* **2009**, *68*, 643–653. [CrossRef]
32. Morse, S.; Vogiatzakis, I.N.; Griffiths, G.H. Space and sustainability: Potential for landscape as a spatial unit for assessing sustainability. *Sustain. Dev.* **2011**, *19*, 30–48. [CrossRef]
33. Haines-Young, R.; Martin, J.; Tantram, D. *Countryside Quality Counts: Tracking Change in the English Countryside. Constructing an Indicator of Change in Countryside Quality*; Report for the Countryside Agency, Defra, English Heritage and English Nature; Nottingham University Consultants Limited: Nottingham, UK, 2004.
34. Luck, G.W.; Harrington, R.; Harrison, P.A.; Kremen, C.; Berry, P.M.; Bugter, R.; Dawson, T.P.; de Bello, F.; Díaz, S.; Feld, C.K.; et al. Quantifying the Contribution of Organisms to the Provision of Ecosystem Services. *BioScience* **2009**, *59*, 223–235. [CrossRef]
35. Syrbe, R.-U.; Walz, U. Spatial indicators for the assessment of ecosystem services: Providing, benefiting and connecting areas and landscape metrics. *Ecol. Indic.* **2012**, *21*, 80–88. [CrossRef]
36. Chan, K.M.; Shaw, M.R.; Cameron, D.R.; Underwood, E.C.; Daily, G.C. Conservation Planning for Ecosystem Services. *PLoS Biol.* **2006**, *4*, e379. [CrossRef] [PubMed]

land

MDPI

Article

Investigating the Diversity and Variability of Eastern Mediterranean Landscapes

Ioannis N. Vogiatzakis * and Paraskevi Manolaki

School of Pure & Applied Sciences, Open University of Cyprus, PO Box 12794, 2252 Nicosia, Cyprus; paraskevi.manolaki@ouc.ac.cy

* Correspondence: ioannis.vogiatzakis@ouc.ac.cy; Tel.: +357-22411933

Received: 31 August 2017; Accepted: 17 October 2017; Published: 20 October 2017

Abstract: The aim of the paper is to examine the variability of eastern Mediterranean landscapes using a common mapping framework relying on Landscape Character Mapping (LCM). LCM was adapted to the region's specificities placing emphasis on the area's coastal nature, landform variation, land use, in particular pastoral tradition, and settlement patterns, an important output of this study. We selected six study areas, in four countries namely Cyprus, Greece, Jordan and Lebanon, based on their rich cultural and natural heritage, covering a NW to SE gradient of both environmental and cultural settings. We used commonly employed landscape metrics to quantify landscape diversity in the study areas. Similarity in landscape types among study area was measured using Sørensen similarity index. The Kruskall–Walis test was used to test the variability among countries in terms of landscape character variation due to physical and cultural factors. Linear regression was used to assess whether landscape diversity increases with area size. The work has identified and mapped a total of 69 landscape types, of which 18 are rare. Rare landscape types were related to specific geomorphology or intensive anthropogenic activities, which do not occur elsewhere in the East Mediterranean region. The highest similarity was recorded between islands and between mountainous areas. The larger the area the higher is its landscape diversity. This works fills a gap in Mediterranean and sets a benchmark standard for landscape characterization work in the East Mediterranean, so as to enable much greater consistency between countries in future landscape mapping exercises and, ultimately, facilitate trans-boundary cooperation in landscape-scale nature and culture conservation.

Keywords: classification; GIS; LCA; Land Description Units; mapping; planning; typology

---

1. Introduction

Landscape diversity is considered the fourth form of diversity [1] which, in landscape ecology, equates to heterogeneity, and underpins much of the work in that discipline [2]. According to PEBDLS "*landscape diversity is the formal expression of the numerous relations existing in a given period between the individual or a society and a topographically defined territory, the appearance of which is the result of the action, over time, of natural and human factors and a combination of both*" [3]. This definition provides the foundation for the explanation of landscape diversity of the Mediterranean Basin (MB), a coherent geographical region, which includes parts of three continents Europe, Asia and Africa. In the Mediterranean the diverse climate, geology and topography and the intense and long human footprint has resulted in a rich mosaic of cultural landscapes that characterizes so much of the basin [4–6].

Due to the common underlying factors, which have shaped its landscapes, the MB is often considered a relatively homogeneous area. Although it is true to a certain extent that many landscapes of common character are present, there are also many others, which are very different, often as a result of extreme gradients (natural or anthropogenic) which can be recognized in the area. The perceived overall uniformity of the area holds well and has been demonstrated for biological and cultural diversity [5–8] but has never really been quantified for landscape diversity. Many of these landscapes are so typical throughout the Mediterranean to the extent that inhabitants and visitors alike can resonate with their distinct character.

The landscapes of the Mediterranean Basin (MB) also experience common pressures, such as rapid change due to urbanisation, rural depopulation, decreased rainfall, increased fire frequency, tourism expansion, social inequalities, political instability, etc. Historically, the area has demonstrated a tendency to adaptation to new socioeconomic and environmental challenges and as such, a great variety of landscapes has been created, but the rate of change may now exceed their resilience. Thus, there is a need to develop appropriate tools to assess and monitor change and to make reasonable judgment about the condition of the landscape and its capacity to change.

The emergence of landscape on the political agenda (e.g., [9–11]) culminated with the European Landscape Convention (ELC) [12]. The ELC calls for the identification of distinct landscape types within a country, assessment of landscapes, understanding landscape change, and develop landscape quality objectives in partnership with stakeholders [13]. Landscape Characterization is an environmental stratification [14] which attempts to summarize environmental variation in the landscape, but is quite distinct to existing stratification schemes since it generally relies on the identification of meaningful landscape units recognizable in the field rather than arbitrary grid cells (but see [15]). By now there is more than one way to carry out Landscape Character Assessment (LCA) in Europe, each of which comes with its own tools and trade-offs [16,17]. LCA has become more complex and holistic over the years with methods ranging from simple–interpretative or mechanistic–analytical to more complex analytical and/or interactive methods [18,19]. These usually lead in a hierarchical system, which allows for upscaling, and downscaling and can be applied in environmental monitoring and policy development [15,17].

In this study, we ask how diverse the landscapes in the Eastern Mediterranean (EastMed) region are, and to what extent the landscapes encountered, are similar among the EastMed countries. In order to do that, we relied on a common mapping framework developed under the MEDSCAPES project, based on the principles of Landscape Character Assessment (LCA) employed in northern Europe. MEDSCAPES, a 2-year project funded by the ENPI-CBCMED program, aimed at the development of LCA as a tool for effective conservation of natural heritage in the Eastern Mediterranean. The project's specific objectives were (a) to develop and apply a best-practice methodology for identifying, mapping and assessing Landscape Character in pilot areas, and (b) to promote the results as a tool for sustainable land use decision-making and landscape-scale protection of the natural and cultural heritage in the East Mediterranean context.

Despite the fact that by now most of the Euro-Mediterranean countries (e.g., Portugal Spain, Italy, Slovenia) have produced a national landscape typology, these are not directly comparable since they all use very different mapping frameworks and often different philosophies to delineate landscapes (see review [20]). At the same time, and while some regional classifications account, to a certain extent, for Mediterranean landscapes [15,21], they rely on generalized patterns often of solely biophysical variables ignoring Mediterranean specificities. None of these have attempted an account of the variation of Mediterranean landscapes at the regional level i.e., they have not been developed explicitly for the Mediterranean region, with the exception of a recent study on land systems [22].

The idea of the present study is to take MEDSCAPES project results a step further and to test the potential of a common landscape mapping framework for assessing and comparing landscape diversity between different countries in the East Mediterranean area. The importance of such work is that it will set a benchmark standard for landscape characterization in the East Mediterranean, so as to enable greater consistency between territories of the Basin in future landscape mapping exercises and, ultimately, facilitate trans-boundary cooperation in landscape-scale nature and culture conservation.

2. Materials and Methods

2.1. *Study Area-East Mediterranean Basin*

Six study areas in four different east Mediterranean countries were selected (Table 1; Figure 1). Their selection was based on their rich cultural and natural heritage, to represent a biogeographical gradient (NW to SE), but also a gradient of both environmental and cultural settings. Therefore, it encompasses both islands and continental areas, including mountains, arid and semi-arid areas, but also the imprints of Muslim and Christian traditions, as well as archaeological evidence of past civilizations, all typical of the Mediterranean character:

(a) Cyprus: the study area includes the whole of the island, the third largest in the Mediterranean. The island is divided into three geomorphological zones, the Troodos Mountain, the Pentadaktylos Range and the Mesaoria plain with rugged morphology and varied geology. The climate is, in general at the drier end of the Mediterranean-type climates. The rural landscape is dominant and usually intermixed with natural elements.

(b) Mujib (Jordan): the study area is located within Madaba and Kerak governorates. The area extends from the Jordan Rift Valley (JRV) escarpment in the west at 420 m below sea level to the central highlands plateau in the east with elevations exceeding 700 m above sea level.

(c) Al Yarmouk (Jordan): the study area is located in the NW part of Jordan. Al Yarmouk catchment area has high relief topography with elevations from about 26 m.b.s.l. to about 1200 m.a.s.l. The catchment comprises of large urban centers (Irbid and Al Ramtha) in addition to extensive agricultural activities (around 50% of the total catchment area).

(d) Epirus (Greece): the study area is predominantly mountainous characterized by significant landscape diversity. High mountains with dense forest cover below 1700 m and alpine meadows at their peaks are traversed by wide cultivated valleys and steep unsettled gorges; cultivated, settled areas and extensive grazed pastures.

(e) Lesvos (Greece): the study area includes the whole of the island, the third largest in the Aegean Sea. Agriculture and tourism are the main economic activities on the island. Approximately 41.3% of the island's surface is olive plantations, 34% maquis and garrigue, 17% forest, 4.3% other crops, while the remaining land has various uses such as constructions, wetlands, etc.

(f) Lebanon: the study area is approximately 4747.40 km^2, c.45% of the total area of the country. It contains 15 cazas that reach an altitudinal range of 1700 m, with exceptions of high peaks in mountainous areas, and low altitudes in deep valleys. Moreover, it encompasses four distinct geomorphological regions, which are: coastal plain, Lebanon mountain range, Bekaa valley, and the Anti-Lebanon mountain range.

Table 1. Main characteristics of the study sites and the datasets used in Landscape Character Mapping.

Natural and Cultural Features of the Study Areas	Area (km^2)	Sources of Information for the Desk Study			
		Geology	Landform	Landuse—Land Cover	Settlement
Cyprus: the islands is a global biodiversity hotspot and is located on a bird migratory route. There are 38 terrestrial Sites of Conservation Importance at the EU level (i.e., Natura 2000 sites) while c.60% of its territory comprises high nature value farmlands important for biodiversity support. Cyprus is also rich archaeological sites as well as religious sites (Christian and Muslim).	9251	Geological map (1:250,000)	20 m contours—derived from USGS Global (DEM)	Corine	- Village Settlement Data: villages clusters - Village Settlement Data: village compactness - Field Patterns
Mujib (Jordan): As part of the Jordan, Rift Valley (JRV) the area in along an important global bird migration flyway. It is one of the most remarkable nature and biodiversity areas of interest in Jordan, with one Protected Area and two Special Conservation Areas The area is also known for its cultural and religious value, and is one of the most visited adventure and therapeutic tourism destinations in Jordan. Large part of the Area is a UNESCO Man and the Biosphere Reserve	4207	Geological map of Jordan (1:250,000)	Topographic maps (1:50.000)	National Land Cover Map 1:250,000	Google Earth Topographic Maps
Al Yarmouk (Jordan): The area includes a Special Conservation Area, a forest reserve while it is also rich in scenic, historical, and cultural elements, including remains of uninterrupted occupation since the Neolithic Period and the ancient city of Gadara (Umm Quais)	1390	Geological map of Jordan (1:250,000)	30 m DEM (ASTER) Global Digital Elevation Model ver. 2; 30 m contours	National Land Cover Map 1:250,000	Google Earth Topographic Maps
Epirus (Greece): The rich natural heritage of the area, which includes the National Parks of Northern Pindos and Vikos-Aoos (both important nationally for their floristic diversity), is complemented by its significant historic and cultural heritage, characteristically portrayed through the plethora of traditional settlements, stone bridges and footpaths. Vikos-Aoos is a UNESCO Global Geopark and one of the last European stongholds of the brown bear	3908	Geological map of Epirus	Topographic map, contours per 50 m (Aster Global DEM) 20 m)	CORINE land-cover (1:250,000), geodata.gov	Google Earth Topographic Maps

Table 1. *Cont.*

Natural and Cultural Features of the Study Areas	Area (km^2)	Sources of Information for the Desk Study			
		Geology	Landform	Landuse—Land Cover	Settlement
Lesvos (Greece): Contains 34% surface cover of maquis and garrigue, 17% forest, and important wetlands. The architecture is particularly rich in legacy and styles and showcases cultural influences from the East (Ottoman Empire) and from the West (Europe). Natural and other monuments and historical sites: The Petrified Forest of Sigri and cultural monuments (i.e., the Castle of Sigri). Religious monuments: the Ypsilou Monastery and the Monastery of Pithariou. Archaeological Sites: Fortress of Sigri, Eresos Acropolis, Monastery High, Pithariou Monastery. Traditional settlements: Vatoussa	1795	Geological map of Lesvos (1:250,000), UoA	Topographic map, contours per 50 m (1:250,000) UoA	CORINE land-cover (1:250,000), geodata.gov	Ordnance survey map (1:250,000); geodata.gov
Lebanon: the study area includes three Man and the Biosphere Reserves, more than 10 nationally designated protected areas and biodiversity hotspots such as the Lebanon and Anti-Lebanon mountains, as well as Byblos (Jbeil) a World Heritage Site	4757	Geological map	50 m contour maps; Topographic map	Land cover map (1:20,000)	Settlement map

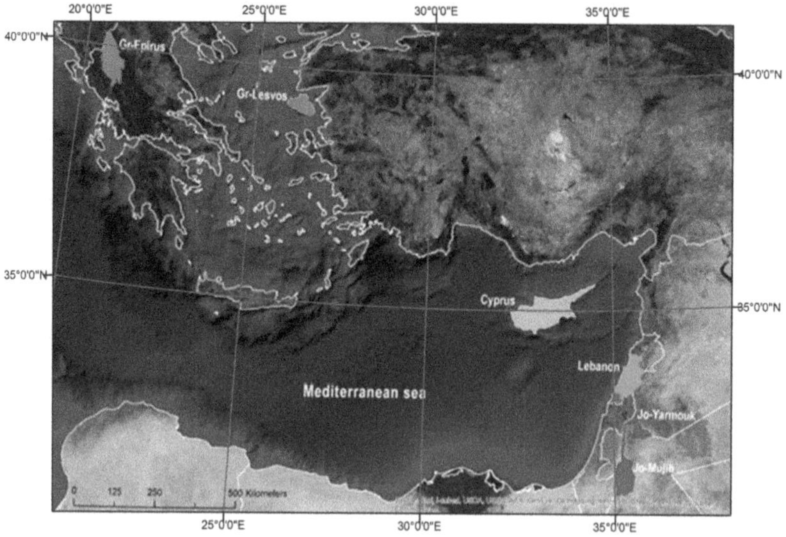

Figure 1. Location of the six study areas, different color indicates study areas located in different countries (from west: Orange = Greece; Yellow = Cyprus; Green = Lebanon and Red = Jordan).

2.2. *Landscape Character Mapping*

The study involved review and evaluation of the LCA methodology in the context of its original principles and objectives as developed and applied in the U.K. The evaluation considered how improvements and additions in the steps involved in UK_LCA methodology will increase subjectivity in the process and better measure the heterogeneity of the eastern Mediterranean landscapes, as described below (Figure 2).

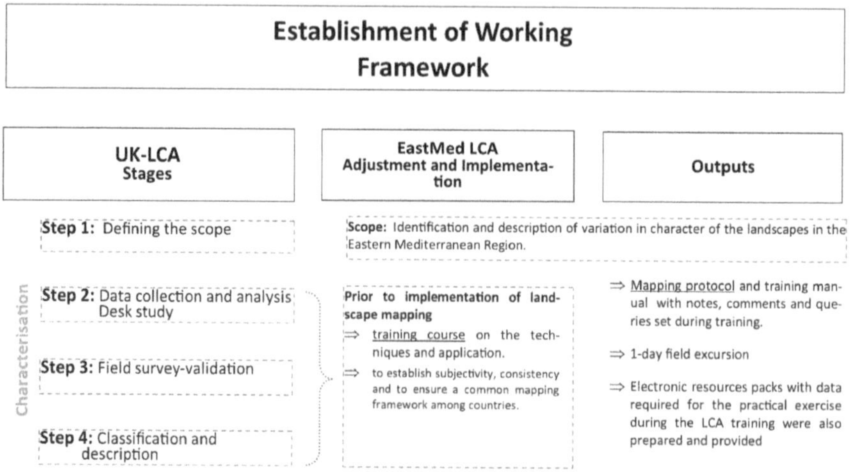

Figure 2. Common Working Framework for Landscape Character Mapping (LCM). Four day training was provided to all teams covering the basics of LCA process including GIS mapping, field validation and evaluation for decision-making and planning.

Mapping procedure: During the implementation, the size of the digitized polygon and which landscape features considered as mappable (mapping unit), was among the most important issues arose during the mapping process. This was overcome by setting the minimum mapping unit (MMU) to 5 km^2. Mapping was carried out at the 1:250,000 scale and successive GIS maps overlay of four definite variables (Table 1) was carried out for all study areas in order to produce Land Description Units (LDUs). LDUs are the largest homogeneous map unit sharing similar pattern of natural and cultural elements and they comprise the basis for further judgment and decision-making [16]. The procedure starts by dividing the area of interest into physiographic units from contour and geological data. More specifically a topographic map (contour lines of 20 m or 50 m) is added in a GIS followed by a simplified geological map. During onscreen overlay the user defines significant relief variations based on the significant changes of form (distance of contour lines and variation of line shapes) and then delineates new polygons which integrate major geology changes in order to define landform units (combination of contour lines and geology). The resulting units are then further sub-divided by land cover and finally by settlement patterns to derive the LDUs. In every stage the unit is populated by the relevant value or class of each key attribute used. For example: unsettled, densely, medium or sparsely settled in the case of settlements. These units were subsequently amalgamated into Landscape Types (also termed Landscape Character Types - LCTs) with similar physical and cultural attributes [16]. Desk study in a GIS is a core part the landscape character mapping; in this case in an attempt to increased accuracy and speed in gathering landscape information we used Google Earth for reconnaissance purposes prior to fieldwork and for correcting LDUs boundaries following fieldwork. In addition, Google Earth imagery served for identifying current land use/land cover (LULC) thus allowing the refinement of LDUs, where up-to-date mapped data was not available.

Field Validation: Field survey was then carried out in order to validate all the information collated during desktop mapping and to confirm that LDUs corresponded to their mapped definition both thematically and in terms of actual boundaries. A standard field sheet was used in order to make and record observations in a systematic and consistent manner. This included information about the physical aesthetic and perceptual characteristics as well as a brief written description, which captured the overall impression of landscape character.

Typology: The developed typology i.e., nomenclature accounted for important elements, which characterize the Mediterranean landscapes mainly summarized in the following factors:

(A) Physical attributes: (i) geomorphology i.e., the inclusion of rugged terrain of hills and mountains, ravines and gorges, as well as plateaus was seen a necessary part of the typology (ii) coasts: MB has a long coast line while 3 out of 6 study areas border the sea, (iii) local distinctness: Fleshing out distinct types present at the national or case study level as described in Section 3.1.

(B) Cultural attributes: (i) Settlement pattern mainly reflected by villages and field pattern (ii) Landuse and land cover: a long history of human activity imprinted on land cover. For example the rangelands, reflecting the pastoral tradition [22–24] and shrublands (i.e., maquis or garrigue) also characteristic of the MB were incorporated.

Landscape Character Types (LCTs) from all countries were entered into a common database. Then, LCTs were subcategorized based on the prominent landform, LULC and settlement pattern (unsettled or settled). Afterwards, based on the description of each LCT an interactive, dynamic process took place, to identify similarities and repetitive patterns among national types. As a general rule, we used the Landform-Land-Use-Settlement pattern to create a broad Mediterranean typology, encompassing the various landscape types identified at Level 1 mapping of the pilot areas. The common EastMed Landscape Typology as derived from the national typologies is showed in the Appendix A.

2.3. Landscape Complexity and Diversity

In order to study the complexity of the study areas in terms of landscape types we employed five commonly employed landscape metrics for diversity, composition and configuration [2] namely LCT

Richness (LCTR); Shannon's diversity index (SHDI), Shannon's evenness index (SHEI), Dominance, Mean Shape index (MSI). Indices were calculated at the landscape level [25] using V-Late 2.0 beta extension for GIS [26] and were chosen since the may be used for comparison between landscapes of different sizes.

2.4. Statistical Analysis

The Sørensen similarity coefficient Ss was employed in order to measure landscape types' similarity between study area pairs [27]:

$$S_s = \frac{2a}{2a + b + c} \quad (1)$$

where, a = the number of common landscape types between two areas, b = the number of landscape types unique to the first of the two areas and c = the number of landscape types unique to the second area.

Also, in order to test if there are significant differences in the landscape variation among countries, and if those differences are due to their physical or cultural differentiation we first grouped the LDUs recorded in the six study areas in ten broader landform categories as well as in six broader land use categories. Then we applied the non-parametric K-independent Kruskal–Wallis test following a test for normality (Smirnov–Kolgomorov–Smirnov $p < 0.05$) for the physical and cultural categories, taking into account the number of LDUs included in each category by country. We also run (log to log) linear regressions to examine the effect of area size on landscape diversity (response variable number of LDUs, number of types, SHFI and SHEI.

3. Results

3.1. East Mediterranean Typology

The highest similarities were between Cyprus and Lesvos, the two sites in Jordan (Mujib and Al Yarmouk), Lebanon and Epirus as well as Lebanon and the two sites in Jordan (SI > 0.4). The lowest similarities between Cyprus and the two sites in Jordan (Table 2).

Table 2. Landscape Types similarity among study sites based on Sørensen similarity coefficient.

	Cyprus	Al Yarmouk	Mujib	Lebanon	Lesvos	Epirus
Cyprus						
Al Yarmouk	0.09					
Mujib	0.04	0.46				
Lebanon	0.22	0.46	0.31			
Lesvos	0.46	0.13	0.12	0.17		
Epirus	0.20	0.34	0.23	0.42	0.16	

The landform character of landscapes in Cyprus and Lesvos is predominantly hilly, whereas in Epirus and Lebanon predominantly upland. The Jordanian study areas are dominated by plateaus. In terms of land use, the landscapes of Cyprus, Lesvos and Al Yarmouk are predominantly of agricultural character, in Mujib characterized by the extent of rangelands, while Lebanon and Epirus are predominantly of forested character (Figure 3).

A total number of 617 LDUs for a total of 69 landscape types at the East Mediterranean level was derived. The resulting East Mediterranean Landscape Typology consists of ten landscape classes based on prominent landform and six classes based on land use (Table 3; Figure 3). The predominant landform in the typology is hills (22% of the types) followed by uplands (20% of the types) and mountains (13% of the types). The predominant land use in the typology is agriculture (46% of the types) followed by forest (24% of the types) and rangelands (19% of the types). There are also 18 types which are unique to the areas covered by this study.

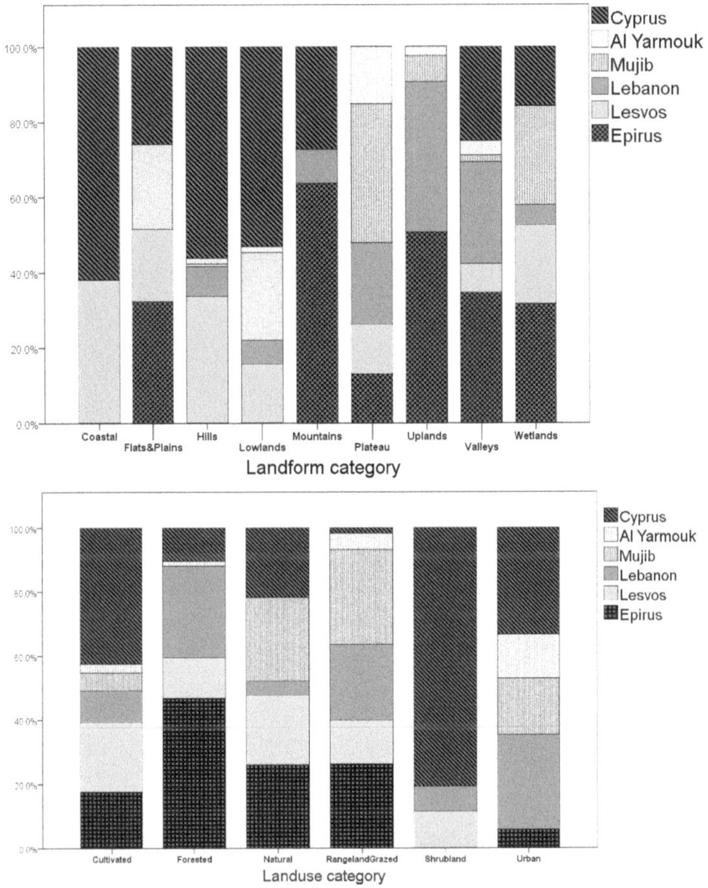

Figure 3. Landscape Profile of the study areas in terms of landuse and landform.

Table 3. Summary of the Eastern Mediterranean Typology.

Total number of LDUs	617
Total number of LCTs	69
Classes (categories) based on landform (Total number)	Hills, Valleys, Mountains, Uplands, Plateau, Coastal, Lowlands, Flats and Plains, Wetlands, Urban (10)
Classes (categories) based on land use (Total number)	Rangelands and Grazed, Shrubland, Forested, Cultivated, Vineyards, Arid (6)
Most common category based on landform (No of LDUs)	Hills (137)
Most common category based on land use (No of LDUs)	Urban (45), Shrubland (41)
Most common LCT	Settled cultivated/agricultural lowlands
Rare types (No of LDUs)	Escarpments of the Dead Sea (1), Dead Sea (1), Salt flats (3), Water bodies (1), River beds (4), Forested sparsely settled lowlands (1), Unsettled Forested Plateaus (1), Settled Forested plateaus (1), Eroded plateau mixed farming (1), Settled High (Alpine) Mountains (1), Unsettled mountain shrublands (1), Settled cultivated coastal valley (1), Coastal dunes (1), Settled rangelands valley (1), Forested sparsely settled lowlands (1), Unsettled lowland shrublands (1), Unsettled cultivated alluvial plains (1), Unsettled plain rangelands (1).

3.2. *Landscape Diversity*

Cyprus with the largest studied area resulted in more mapped Land Description Units (210 in total) and consequently more landscape types (35). All studies areas are diverse, with Lebanon having the highest SHDI value (0.901). Dominance was higher in Cyprus (0.838) and lower in Lesvos (0.246) with the rest of the study areas having intermediate values. The LDUs in Lebanon, Yarmouk and Mujib study areas are the most complex in terms of Mean Shape Index (>2.2) (Table 4).

Table 4. Landscape Diversity in the six study areas.

Area Analysis	Lesvos	Epirus	Mujib	Al Yarmouk	Lebanon	Cyprus
Total Area in km^2	1795	3908	3800	1390	4757	9251
Number of LDUs	95	151	55	21	97	198
	Diversity Analysis					
Landscape Character Type Richness (LCTR)	11	16	23	16	21	35
Shannon's Diversity (SHDI)	2.152	2.28	2.601	2.321	2.742	2.717
Shannon's Evenness (SHEI)	0.897	0.822	0.83	0.837	0.901	0.764
Dominance	0.246	0.493	0.534	0.452	0.303	0.838
	Form Analysis					
Mean Shape Index (MSI)	1.95	1.747	2.215	2.203	2.263	2.073

Statistical analysis results indicated that, apart from Mujib and Lesvos, for all other sites there was a statistical significant difference in the number of LDUs within the main land cover categories and the number of LDUs within the main landform categories (Table 5). In addition linear regressions showed that although most relationships between area and the components of diversity have medium to high predictive power only the relationship between area and landscape type richness was significant (R2 = 0.73, $p < 0.05$; Table 5).

Table 5. Results for the statistical tests performed (K-Independent test between Land Description Units –LDUs, per site and Landform categories).

	Cyprus	Al Yarmouk	Mujib	Lebanon	Lesvos	Epirus
	K-Independent test between LDUs per site and Landform categories					
Chi-Square	32.159	19.194	15.613	28.741	10.881	33.223
df	9	9	9	9	9	9
Asymp. Sig.	0.000	0.024	0.075	0.001	0.284	0.000
	K-Independent test between LDUs per site and Landuse categories					
Chi-Square	25.578	17.659	18.620	12.416	10.566	14.598
df	5	5	5	5	5	5
Asymp. Sig.	0.000	0.003	0.002	0.030	0.061	0.012
	Linear Regressions (log to log)			R2	F	P
	LDUs against area			0.55	4.70	0.08
	Landscape types against area			0.73	10.96	0.03 *
	SHDI against area			0.61	6.33	0.07
	SHEI against area			0.25	1.31	0.31

* statistically significant at the 0.05 level.

4. Discussion

The study has filled a gap in East Mediterranean landscape mapping and typology by setting a benchmark for landscape characterization work in the area, while at the same time, recognized and revealed the distinct landscape character of each study area and the inherent diversity of its landscapes. The typology produced resulted in sufficient variation to describe landscape types while it emphasized elements which are characteristic of Mediterranean landscapes. More specifically it has accounted for the coastal nature of the area, the variety of landforms and has incorporate settlement pattern and land use, in particular pastoral tradition the latter being an important element of Mediterranean land systems [22].

Although the pace of landscape characterization and mapping has been slower in the Mediterranean compared to Northern Europe there has been significant progress in recent years [20,28]). A comparison with the existing typologies which account for Mediterranean landscapes [15,21] revealed that the first study identified 7 landscape types in the whole Mediterranean Basin while the second 11 landscape types at the 3rd level of classification. Thematically, although there is some correspondence, these are neither detailed nor eastern Mediterranean enough, to capture common as well as rare landscape types of the region as identified by the study. This is mainly due to the coarse mapping scale but also their geographical extent and scope (pan-European). For example, the extensive coastline is usually neglected in many regional and national land based typologies. This has led to the development of Seascapes Assessment and a call for treating islands as separate entities [28]. Rangelands are important elements of the landscapes, culturally and economically, in the Mediterranean and the study areas examined are no exception. Grazing may be year-round or seasonal (transhumance)—or even more sporadic in semi-desert areas. Low vegetation, often with small, spiny shrubs in combination with grasses and herbs have often been mapped as "bare land" in coarse land-use maps with the exception of recent work on Mediterranean land systems [22] in which 27 major land systems, not landscapes, were identified.

The high diversity of landscapes highlighted by our study is expected given the natural and cultural diversity of the study area. The work has identified and mapped a total of 69 landscape types (69) of which 18 rare. Rare landscape types (Table 3) are related to specific geomorphology or intensive anthropogenic activities which do not occur elsewhere in the region. The analysis of the landscape categories revealed statistically significant differences among countries apart from the landscape types with predominant landform of "*Flats and Plains*", which they were almost equally distributed among study sites. However, they were absent from Lebanon and Al Yarmouk. In addition, "*Coastal*" landscape types were present only in the cases of the two islands (Cyprus and Lesvos). "*Mountainous*" landscape types were found only in three study sites Cyprus, Lebanon and Epirus with the latter to have the highest percentage. In terms of land use categories, Shrubland is by far the most common landscape type in Cyprus; however, it was observed only in three study sites (Cyprus, Lebanon and Lesvos). All the other land use categories were found in all study sites with cultivated landscapes to be more abundant in Cyprus, forested landscapes in Epirus and grazed lands in Mujib.

Each study site shares common landscape types derived from common environmental and long human history in the area. The highest similarities were observed on the islands (Cyprus and Lesvos) and the mountainous areas (Lebanon and Epirus), which were then followed by the similarity of the latter to Cyprus, due to the mountainous character of the island itself. However, within each study site, distinctive LCTs were derived, which were originated from the particular conditions prevailing in each country and involve both physical such as "Arid" or "Dead Sea" special character types and anthropogenic ones like "Artificial settled hills" in Lebanon.

While Lesvos and Al Yarmouk are comparable in area, they demonstrated a large divergence in their number of LDUs. This suggests that LDUs are generally larger in Yarmouk and more irregularly shaped as indicated by the higher MSI value. Although smaller, Yarmouk has higher LCT richness unevenly distributed as indicated by the higher dominance value. Mujib is comparable to Epirus and Lebanon study areas, in terms of area extent with fewer LDUs and the highest LCT richness among the three. This suggests that LDUs are larger and irregularly shaped (as indicated by the MSI) while the distribution of area among LCTs is uneven, i.e., a few LCTs are dominant. Cyprus, the largest in extent study area, has the largest number of LDUs and the highest richness of LCTs. The island demonstrates the highest dominance value (ergo the lowest SHEI) compared to the rest study areas suggesting that the distribution among different LCTs is more uneven and therefore a few LCTs dominate the landscape in Cyprus. The lower dominance values in Lesvos and Lebanon indicate that LCTs cover equal surfaces in these case studies. Complexity of a landscape as expressed by its shape (quantified by MSI) reflects boundaries between semi-natural areas, but also management in cultural landscapes. Simple shapes is the norm for agricultural and pastoral landscapes while on mountain or other landscapes dominated by landform or geology boundary types might be more abrupt. In our case, MSI values among study sites are very close demonstrating similarity on the range and types of boundaries present.

Regression showed that landscape diversity (expressed by the number of landscape types and SHDI) is area dependent i.e., the largest the area the highest the landscape diversity. The relationship seems to be analogous to the species-area curve [29]. However, when examining evenness against area size there is no trend since SHEI measures the distribution of area among patch i.e., in this case LDU types. The use of Shannon index for assessing landscape diversity is recommended since it is sensitive to rare landscape types and therefore appropriate for landscape management within an ecological framework [30]. The highest number of landscape types recorded in Cyprus compared to the rest of the study areas should not be attributed only to its size but also to the fact that we mapped the whole range of landscapes within the island's geographical boundaries. Landscape diversity comparisons among regions depend usually on the landscape itself, but also on the way that the methodology has been applied in each region. Despite the effort to streamline the process we cannot exclude elements of subjectivity in the application of the landscape character mapping by the teams involved. In terms of characterization, the findings imply that the 'complexity' of the classification scheme (should) increase as the area increases.

The problems of cross comparison between countries has been addressed by other studies [13,17]. In the absence of truly unified datasets for the whole area the study relied on nationally available datasets which differed in nomenclature and classification systems. Data issues (availability, accessibility and quality) along with problematic access to some parts of the study areas due to safety and security reasons was also a limitation in addition to semantics in the interpretation of landform types, settlements, and the very term landscape. However, and despite the diverse background of the persons involved in the study, training and protocol development brought the teams closer than before to a common understanding of methods and terms. In the same way that sampling units assist when comparing population samples, a consistent landscape mapping framework can have a similar role when comparing landscape types.

5. Conclusions

Landscape mapping and identification, a prerequisite for countries signatories to the ELC, is a means to inform landscape planning, develop strategies for landscape conservation and enhancement, or to provide outputs which can be used with other decision-making tools such as Environmental Impact Assessment. For example whereas previously, Greece and Cyprus had recognized a number of outstanding landscapes and landscape features as worthy of protection, they now need to address in their spatial planning all types of landscapes, whether these are outstanding, "every-day" or even degraded and in need of attention. Therefore identifying the variety of landscapes within a country's geographical boundaries is the first step towards landscape oriented planning. The landscape types identified herein form by no means an exhaustive list with the exception of Lesvos and Cyprus, since we did not map the whole range of landscapes within the countries geographical boundaries. The development of the typology requires the sampling of the whole range of landscape units in a geographical area in order to identify the attributes that discriminate between the full complements of landscape types. However, the framework devised herein can be used for a range of applications (ecological, cultural, planning) and has already been adopted by the relevant Planning Departments in Cyprus and Lebanon through their Local/Regional Development Plans. In addition it can assist with infusing ELC principles beyond European boundaries and with the development of a Mediterranean-wide landscape typology as a common language for monitoring landscape changes in a very dynamic and fragile region of the world.

Acknowledgments: This work was partly funded by the MEDSCAPES project (ENPI-CBCMED program). We would like to thank the MEDSCAPES team for providing access to datasets.

Author Contributions: The idea was a product of discussions between the authors. Ioannis Vogiatzakis wrote the manuscript while Paraskevi Manolaki carried out the statistical analysis and contributed with comments to all versions of the manuscript.

Conflicts of Interest: The authors declare no conflict of interest.

Appendix A

Table A1. LEB = Lebanon, ALYAR = Al Yarmouk, CY = Cyprus, LES = Lesvos, MUJ = Mujib, EPIR = Epirus).

		Mediterranean Type	National Types
Hills	Hills1	Settled rangeland hills	Rangeland sparsely settled hills (LEB); The majority are rangelands and cultivated hills while settled hills are minority (ALYAR); Settled rangeland hills (CY)
	Hills2	Unsettled forested hills	Forested hills unsettled (LES); Unsettled forested hills (CY)
	Hills3	Settled forested hills	Forested settled hills (LEB); Settled forested hills (CY)
	Hills4	Settled cultivated hills	Cultivated hills sparsely settled (LES); Settled cultivated hills (CY)
	Hills5	Unsettled hill shrublands	Scrubby hills and valleys unsettled (LES); Unsettled hill shrublands (CY)
	Hills6	Unsettled Arid hills sporadically grazed/Arid hills	Arid hills, valleys and lowlands sporadically grazed (or by wild animals) (MUJ); The majority are arid hills while cultivated hills are minority (ALYAR)
	Hills7	Unsettled cultivated hills	Unsettled cultivated hills (CY)
	Hills8	Settled hill shrublands	Settled hills shrublands (CY)
	Hills9	Settled hill with vineyards	Settled hills with vineyards (CY)
	Hills10	Unsettled rangeland hills	Grazed hills and valleys unsettled (LES)
Uplands	Uplands1	Unsettled forested uplands	Forested Uplands (EPIR); Forested unsettled high uplands (LEB); Forested uplands (Unsettled) (ALYAR)
	Uplands2	Settled forested uplands	Forested Uplands (EPIR); Forested settled uplands (LEB); The majority are forested uplands while settled uplands are minority(Settled forested uplands) (ALYAR)
	Uplands3	Settled cultivated uplands	Cultivated Uplands (EPIR); Agricultural settled high uplands (LEB); Agricultural settled uplands (LEB); Sparsely settled, grazed and cultivated uplands (MUJ); The majority are cultivated uplands while settled uplands are minority (ALYAR)
	Uplands4	Unsettled cultivated uplands	Cultivated Uplands (EPIR);
	Uplands5	Settled shrubland uplands	Scrubby sparsely settled high uplands (LEB);
	Uplands6	Unsettled rangeland uplands	Grazed High Uplands (EPIR); Grazed Uplands and Hills (EPIR); Rangelands unsettled high uplands (LEB); Unsettled upland rangeland (MUJ)
	Uplands7	Settled rangeland uplands	Grazed High Uplands (EPIR); Grazed Uplands and Hills (EPIR); Rangelands sparsely settled uplands (LEB)

Table A1. *Cont.*

		Mediterranean Type	National Types
Plateau	Plateau1	Unsettled alpine plateaus	Alpine Plateaus (EPIR)
	Plateau2	Unsettled forested plateaus	Forested plateaus (LES)
	Plateau3	Settled forested plateaus	Forested plateaus (LES)
	Plateau4	Settled cultivated plateaus	Cultivated Plateaus and Hills (EPIR); Cultivated plateau (LES); Agricultural sparsely settled high plateau (LEB); Agricultural sparsely settled plateau (LEB);Cultivated and rangelands plateaus (settled cultivated plateaus) (ALYAR); Cultivated plateau (settled cultivated plateaus) (ALYAR); Settled and cultivated plateaus (ALYAR); The majority are rangelands and cultivated plateau while settled hills are minority (ALYAR); Settled plateaus, cultivated and mixed agriculture-rangeland (MUJ)
	Plateau5	Unsettled cultivated plateaus	Cultivated Plateaus and Hills (EPIR); Cultivated plateau (LES);
	Plateau6	Unsettled rangeland semi-arid/eroded plateaus	Rangeland unsettled high plateau (LEB); Semi-arid eroded plateau rangelands (MUJ); Semi-arid plateau rangelands (MUJ); Eroded plateau mixed farming and rangeland (MUJ);
	Plateau7	Settled rangeland semi-arid/eroded plateaus	Semi-arid eroded plateau rangelands (MUJ); Semi-arid plateau rangelands (MUJ); Eroded plateau mixed farming and rangeland (MUJ); Rangelands and cultivated plateau (ALYAR)
	Plateau8	Eroded plateau mixed farming	Eroded plateau mixed farming (ALYAR)
Mountains	Mountain1	Settled forested mountains	Settled mountain forest (CY); Forested Mountains (EPIR)
	Mountain2	Unsettled forested mountains	Unsettled mountain forest (CY); Forested Mountains (EPIR)
	Mountain3	Unsettled High (Alpine) mountains	High Mountains (EPIR)
	Mountain4	Settled Alpine mountains	High Mountains (EPIR)
	Mountain5	Settled mountain shrublands	Settled mountain shrublands (CY)
	Mountain6	Unsettled mountain shrublands	Unsettled mountain shrublands (CY)
	Mountain7	Settled cultivated mountains	Settled cultivated mountains (CY)
	Mountain8	Unsettled rangeland mountains	Rangeland unsettled mountainous area (LEB); Grazed Mountains (EPIR)
	Mountain9	Settled rangeland mountains	Grazed Mountains (EPIR)
Coastal	Coastal1	Settled cultivated coastal alluvial plains	Settled cultivated coastal alluvial plains (CY)
	Coastal2	Settled coastal hill shrublands	Settled coastal hill shrublands (CY)
	Coastal3	Unsettled coastal hill shrublands	Unsettled coastal hill shrublands (CY)
	Coastal4	Settled cultivated coastal hills	Settled cultivated coastal hills (CY); Cultivated hills sparsely settled (LES)
	Coastal5	Settled coastal lowland shrublands	Settled coastal lowland shrublands (CY)
	Coastal6	Settled cultivated coastal lowlands	Settled cultivated coastal lowlands (CY); Cultivated hilly lowlands sparsely settled (LES)
	Coastal7	Settled cultivated coastal valley	Settled cultivated coastal valley (CY)
	Coastal8	Unsettled coastal lowland shrublands	Unsettled coastal lowland shrublands (CY)
	Coastal9	Unsettled cultivated coastal lowlands	Unsettled cultivated coastal lowlands (CY); Cultivated hilly lowlands unsettled (LES)
	Coastal10	Coastal dunes	Coastal dunes (CY)
	Coastal11	Unsettled grazed coastal hills	Unsettled Grazed coastal hills (LES)
	Coastal12	Settled cultivated coastal plains	Settled cultivated plains (LES)

Table A1. *Cont.*

		Mediterranean Type	National Types
Valley	Valley1	Settled forested valley	Settled forested valley (CY); Forested Valleys (EPIR); Forested sparsely settled valley (LEB)
	Valley2	Unsettled forested valley	Forested Valleys (EPIR); Forested unsettled deep valleys (LEB)
	Valley3	Settled cultivated valleys	Settled cultivated valley (CY); Cultivated Valleys (EPIR); Cultivated valleys (LES); Agricultural sparsely settled valley (LEB); Cultivated valley floors and river terraces (ALYAR)
	Valley4	Unsettled cultivated valleys	Unsettled cultivated valley (CY); Cultivated Valleys (EPIR); Cultivated valleys (LES)
	Valley5	Unsettled deep valleys	Unsettled ravine shrublands (CY); Deep Ravines (EPIR); Unsettled deep valleys and ravines (MUJ)
	Valley6	Settled valley rangelands	The majority are valley rangelands while the settled valley are minority (ALYAR)
Lowlands	Lowlands1	Forested settled lowlands	Forested sparsely settled lowlands (LEB)
	Coastal5	Unsettled cultivated lowlands	Unsettled cultivated lowlands (CY); Cultivated hilly lowlands unsettled (LES)
	Lowlands3	Settled cultivated lowlands	Settled cultivated lowlands (CY); Agricultural sparsely settled lowlands (LEB); Cultivated hills lowlands sparsely settled (LES)
	Lowlands4	Unsettled lowland rangelands	Rangelands/lowland/unsettled (ALYAR); Unsettled lowland, hill and valley rangelands, regularly/seasonally grazed (MUJ)
	Lowlands5	Settled cultivated alluvial lowlands	Settled cultivated alluvial lowlands (CY)
	Lowlands6	Unsettled lowland shrublands	Unsettled lowland shrublands (CY)
Wetlands	Wetlands1	Coastal wetlands	Coastal Wetlands (CY); Coastal wetlands (LES)
	Wetlands2	Lakes	Lakes (EPIR)
	Wetlands3	Riverbeds	Riverbeds (EPIR)
	Wetlands4	Water bodies	Water bodies (LEB)
	Wetlands5	Salt flats	Salt flats (MUJ)
	Wetlands6	Dead Sea	Dead Sea (MUJ)
	Wetlands7	Escarpments of the Dead Sea	Escarpments of the Dead Sea (MUJ)
Flats & Plains	Flats&Plains1	Settled cultivated plains	Settled cultivated plains (MUJ); Settled cultivated plains (LES); Cultivated Plains (EPIR)
	Flats&Plains2	Unsettled cultivated plains	Cultivated Plains (EPIR)
	Flats&Plains3	Settled cultivated alluvial plains	Settled cultivated alluvial plains (CY)
	Flats&Plains4	Unsettled cultivated alluvial plains	Unsettled cultivated alluvial plains (CY)
	Flats&Plains5	Unsettled rangeland plains	Unsettled plain rangelands (MUJ)
Urban	Urban	Urban	Urban Areas (EPIR); Urban (CY); Settled plateau (ALYAR); Settled uplands (ALYAR); Artificial settled hills (LEB); (Artificial) Urbanized lowlands (LEB)

References

1. Odum, E.P. Great ideas in ecology for the 1990s. *BioScience* **1992**, *42*, 542–545. [CrossRef]
2. Forman, R.; Godron, M. *Landscape Ecology*; John Wiley: New York, NY, USA, 1986.
3. Council of Europe/UNEP & Environmental Centre for Nature Conservation. *The Pan-European Biological and Landscape Diversity Strategy. A Vision for Europe's Natural Heritage*; Council of Europe Press: Strasbourg, France, 1996.
4. Vogiatzakis, I.N.; Pungetti, G.; Mannion, A. (Eds.) *Mediterranean Island Landscapes: Natural and Cultural Approaches*; Landscape Series; Springer Publishing: New York, NY, USA, 2008; Volume 9.
5. Grove, A.T.; Rackham, O. *The Nature of Mediterranean Europe: An Ecological History*; Yale University Press: London, UK, 2001.
6. Blondel, J.; Aronson, J.; Bodiou, J.-Y.; Boeuf, G. *The Mediterranean Region: Biological Diversity in Space and Time*; Oxford University Press: Oxford, UK, 2010.

7. Athanassopoulos, E.F.; Wandsnider, L. (Eds.) *Mediterranean Archaeological LandscapesCurrent Issues*; University of Pennsylvania Press: Philadelphia, PA, USA, 2004.
8. Mazzoleni, S.; Di Pasquale, G.; Mulligan, M.; Di Martino, P.; Rego, F. (Eds.) *Recent Dynamics of the Mediterranean Vegetation and Landscape*; John Wiley & Sons: New York, NY, USA, 2004.
9. IUCN. *Guidelines for Protected Area Management Categories*; CNPPA with the assistance of WCMC; IUCN: Gland, Switzerland; Cambridge, UK, 1994.
10. Vogiatzakis, I.; Papayannis, T.; Mannion, A.M. Political landscapes of Mediterranean islands. In *Mediterranean Island Landscapes*; Vogiatzakis, I.N., Pungetti, G., Mannion, A.M., Eds.; Springer: New York, NY, USA, 2008; pp. 100–114.
11. Stanners, D.; Bourdeau, P. *Europe's Environment: The Dobris Assessment*; European Environment Agency: Copenhagen, Denmark, 1995.
12. European Landscape Convention. Available online: https://rm.coe.int/16800805ce (accessed on 19 October 2017).
13. Washer, D.; Jongman, R. (Eds.) *European Landscapes: Classification, Evaluation and Conservation*; Environment Technical Reports; European Environment Agency: Copenhagen, Denmark, 2003.
14. Jongman, R.H.G.; Bunce, R.G.H.; Metzger, M.J.; Mücher, C.A.; Howard, D.C.; Mateus, V.L. Objectives and applications of a statistical environmental stratification of Europe. *Landsc. Ecol.* **2006**, *21*, 409–419. [CrossRef]
15. Mücher, C.A.; Klijn, J.A.; Wascher, D.M.; Schaminée, J.H.J. A new European Landscape Classification (LANMAP): A transparent, flexible and user-oriented methodology to distinguish landscapes. *Ecol. Indic.* **2010**, *10*, 87–103. [CrossRef]
16. Warnock, S.; Griffiths, G.H. Landscape Characterisation: The Living Landscapes approach in the UK. *Landsc. Res.* **2015**, *40*, 261–278. [CrossRef]
17. Hazeu, G.W.; Metzger, M.J.; Mücher, C.A.; Perez-Soba, M.; Renetzeder, C.H.; Andersen, E. European environmental stratifications and typologies: An overview. *Agric. Ecosyst. Environ.* **2011**, *142*, 29–39. [CrossRef]
18. Brabyn, L. Landscape classification using GIS and national digital databases. *Lands. Res.* **1996**, *21*, 277–300. [CrossRef]
19. Groom, G.; Wascher, D.; Potschin, M.; Haines-Young, R. Landscape character assessments and fellow travellers across Europe: A review. In *Landscape Ecology in the Mediterranean: Inside and Outside Approaches*; Bunce, R.G.H., Jongman, R.H.G., Eds.; IALE Publication Series: Faro, Portugal, 2006; Volume 3, pp. 227–231.
20. Vogiatzakis, I.N. Mediterranean experience and practice in Landscape Character Assessment. *Ecol. Med.* **2011**, *37*, 17–31.
21. Meeus, J.H.A. Pan-European landscapes. *Landsc. Urban Plan.* **1995**, *31*, 57–79. [CrossRef]
22. Malek, Ž.; Verburg, P. Mediterranean land systems: Representing diversity and intensity of complex land systems in a dynamic region. *Landsc. Urban Plan.* **2017**, *165*, 102–116. [CrossRef]
23. Le Houerou, H.N. Impact of man and his grazing animals. In *Ecosystems of the World 11: Mediterranean Type—Shrublands*; Di Castri, F., Goodall, D.W., Sprecht, R.L., Eds.; Elsevier: Amsterdam, Netherlands, 1981.
24. Papanastasis, V.P.; Chouvardas, D. Application of the state-and-transition approach to conservation management of a grazed Mediterranean landscape in Greece. *Israel J. Plant Sci.* **2005**, *53*, 191–202. [CrossRef]
25. Lang, S.; Tiede, D. vLATE Extension für ArcGIS—Vektorbasiertes Tool zur quantitativen Landschaftsstrukturanalyse. Available online: https://sites.google.com/site/largvlate/gis-tools/v-late (accessed on 19 October 2017).
26. McGarigal, K. *Fragstats: Spatial Pattern Analysis Program for Categorical Maps*; User Manual; University of Massachusetts: Amherst, MA, USA, 2015.
27. Kent, M. *Vegetation Description and Data Analysis: A Practical Approach*, 2nd ed.; Wiley: New York, NY, USA, 2012.
28. Vogiatzakis, I.N.; Zomeni, M.; Mannion, A.M. Characterizing Islandscapes: Conceptual and Methodological Challenges Exemplified in the Mediterranean. *Land* **2017**, *6*, 14. [CrossRef]
29. MacArthur, R.H.; Wilson, E.O. *The theory of Island Biogeography*; Princeton University: Princeton, NJ, USA, 1967.
30. Nagendra, H. Opposite trends in response for the Shannon and Simpson indices of landscape diversity. *Appl. Geogr.* **2002**, *22*, 175–186. [CrossRef]

land

MDPI

Article

The Influence of Geology on Landscape Typology in Jordan: Theoretical Understanding and Planning Implications

Taleb Odeh [1,*], Natalia Boulad [2], Omar Abed [2], Anas Abu Yahya [2], Nour Khries [2] and Nizar Abu-Jaber [3]

1 Arid Lands Academy, The Hashemite University, P.O. Box 330127, Zarqa 13133, Jordan
2 The Royal Society for the Conservation of Nature (RSCN), P.O. Box 1215, Jubeiha 11941, Jordan; natalia@rscn.org.jo (N.B.); omar.abed@rscn.org.jo (O.A.); anas.sabbarenie@rscn.org.jo (A.A.Y.); n.khries@yahoo.com (N.K.)
3 Center for the Study of Natural and Cultural Heritage, German Jordanian University, P.O. Box 35247, Amman 11180, Jordan; nizar.abujaber@gju.edu.jo
* Correspondence: talebs@hu.edu.jo or taleb_odeh@yahoo.com

Received: 20 April 2017; Accepted: 26 July 2017; Published: 31 July 2017

Abstract: Landscape Character Assessment (LCA) has been introduced into Jordan through the MEDSCAPES project. The purpose of this project was to streamline landscape studies and integrate them into the land use planning practices in Jordan. Two areas within the Mediterranean and arid climatic zones of the country were chosen as test areas for the methodology. These were the Yarmouk River drainage basin in the northwest of the country and the Mujib River area in the west of Jordan within the Dead Sea basin. Landscape Character Mapping resulted in 22 and 64 Land Description Units (LDUs) for the Yarmouk and Mujib areas, respectively, which were then classified into 14 landscape types. The factors which control the spatial distributions of these units are geology, land cover, landform, and settlements. However, the study suggests that the underlying geology, which influences topography, impacts indirectly on soil types, climate zones, and human activities, and hence has a predominant influence on the character of these units. Specifically, the transition between the Dead Sea Rift Valley and the adjacent highlands create variations in the topographical relief, climate, water availability, and human settlements. Implementation of LCA in Jordan has done much to highlight geological hazards, such as sinkholes, as constraints to development in certain areas. Here, we described how the LCA process could be implemented in Jordan and how this can help in improving land use management practices in the country.

Keywords: comparative study; landforms; Landscape Character Assessment (LCA); Land Description Unit (LDU); spatial distributions; topography

1. Introduction

Over the last 100 years, population growth and urbanization in Jordan has tremendously changed the land cover of central and north-west Jordan, where there are water resources. The rest of Jordan is mostly desert areas with limited population and Human activity [1]. Population growth and similar, albeit more modest, spatial growth has occurred in the other major cities of the country as well [2]. Poor planning practices during rapid growth periods, which were often triggered by migrations from wars in surrounding countries, have led to a myriad of social and environmental problems [1,2]. These include the common problems of urban sprawl and the loss of agricultural land [3]. More recent legislation and practices rely on long-term master plans for the major cities, which has placed more discipline into the growth patterns [4]. On the other hand, problems of poorly regulated growth are also seen in smaller towns and in rural areas as well, and are perhaps even more pronounced there [1,3].

There are a number of other reasons for the poor land use planning practices in Jordan. These include political pressure imposed by local people and a desire to maintain social cohesion at any cost [4,5]. The cost is high, however, in extending services to far flung scattered houses and building the infrastructure to serve them. Partially for this reason, municipalities in Jordan are generally poorly funded and ill equipped to provide basic services to their residents [3–5].

Of course, there are environmental costs as well [1]. These include groundwater depletion and contamination due to the lack of sewage networks, the changing of surface water drainage patterns, the fragmentation of often sensitive ecosystems, and increased soil loss and the deterioration of agricultural lands [1,6].

As part of the effort to remedy this situation, the Ministry of Municipal and Rural Affairs (MoMRA) in Jordan has begun to develop local master plans for many of the 99 municipalities of Jordan [5]. This has coincided with the funding of the MEDSCAPES (Mediterranean Landscapes) project ENPI-CBCMED (European Neighbourhood and Partnership Instrument Cross Border Component Mediterranean) program of the European Union. The purpose of this project was to introduce an integrative landscape character assessment (LCA) framework for enhanced and sustainable territorial planning and decision-making in the Eastern Mediterranean area [7,8]. The MEDSCAPES project had a number of components that are relevant to this introduction. Within it, LCA has been adapted for local conditions and implemented in two test areas (Figure 1). Finally, major stakeholders, particularly MoMRA, have been involved in the entire project and have signed a Memorandum of Understanding to use LCA as part of the base map set that is used for land use planning throughout Jordan.

The first objective of the research is to demonstrate the first application of landscape character assessment in Jordan that was carried out as part of the MEDSCAPES project. The second objective is to evaluate how the underline geological structures control the spatial distribution of the landscape units.

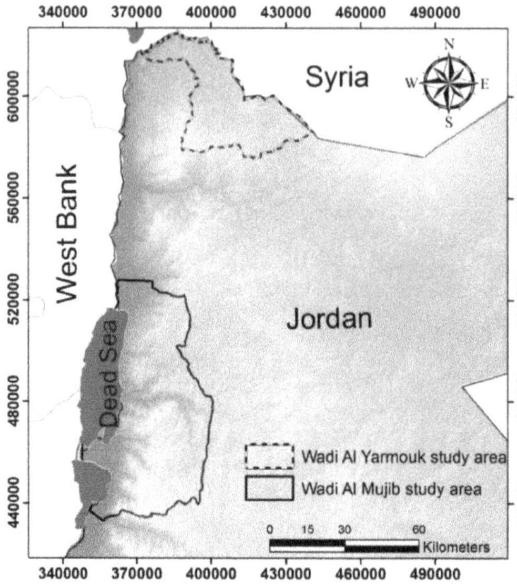

Figure 1. Location of the studied areas. The Wadi Yarmouk study area in the north-west of Jordan which has historical sites, a large area of agricultural activities, and an urbanized area. The Wadi Mujib study area in the Middle West of Jordan which has tourism investments and a beach at the Dead Sea. The names of Yarmouk and Mujib are written with letters (Al) in some references.

2. Background information

The two case studies selected have a mix of natural and cultural landscape characteristics. They are as follows:

2.1. The Yarmouk River Basin

This basin is situated in the northwestern corner of the country and covers an area of about 1160 km^2. It has the forth and tenth largest cities of Jordan; Irbid (population 307,480) and Ramtha (population 74,901). The headwaters of the basin fall in the Ajloun highlands south of Irbid, but the Yarmouk River itself begins permanent flow at Wadi Shallaleh near the Syrian border. The topographic relief in the area is high, ranging from 1200 m above sea level (masl) in the Ajloun Highlands to about −16 (mbsl) at the confluence between the Yarmouk and Jordan Rivers. The geology is dominated by Upper Cretaceous limestone with sinkholes mixed with marl in the lower formations (the Naur and Wadi Shuib formations) and chert in the upper formations (The Amman and Um Rijam Formations). The limestone forms a dome structure with the apex in the area of Ajloun, with the various formations dipping away from this apex in all directions. Towards the east, the Dead Sea Rift Valley is the dominant structural feature, with down rifting leading to rapidly lower elevations in that direction. Towards the north, a peneplain is found at the Yarmouk River showing distinct entrenched meanders along much of the river course. Finally, early in the rifting and uplift process, volcanism led to basalt flows along much of the northern part of this study area [9].

Rainfall falls in the winter, with annual precipitation rates ranging from 580 mm/year in Ras Munif in Ajloun to 390 mm/year in the northern Jordan Valley at Baqoura [10]. The minimum annual temperature ranges from 5–10 °C and mean annual maxima range from 20–30 °C. Soil is dominated by Terra rosa and Rendzina soils [3]. It has significant agricultural activities, ranging from rain-fed olive groves, orchards, and vineyards in the higher elevation to field crops (wheat and rain-fed vegetables in the Irbid area), to irrigated vegetables, citrus, and bananas in the vicinity of the Yarmouk River and the northern Jordan Valley [4]. The area is rich in scenic, historical, and cultural values, including remains of uninterrupted occupation since the Neolithic Period [11].

2.2. Wadi Mujib River Basin

It is situated in Middle West of Jordan with an area of about 3800 km^2. Unlike the Yarmouk study area, Wadi Mujib is not strictly a river drainage system, but a national park stretching from the Mujib River to Sweimeh on the northern shore of the Dead Sea [12]. It extends from the shore of the Dead Sea to the highlands overlooking it. The area has two major cities: Madaba (84,600 inhabitants) and Al Karak (21,678 inhabitants) [5]. The area is important from both an economic as well as an environmental perspective. In addition to the Dead Sea shore, the area has the Ma'in and Zara hot springs as well as beautiful vistas, including a panoramic view of the Dead Sea where a visitor center and small museum known as the Dead Sea Panorama Complex has been built. The area is undergoing rapid development, with large investments in tourism being made on the Dead Sea shore [8]. The Wadi Mujib river basin has very cold winters with temperatures that might go down up to −10 °C in the mountainous area, and very hot, dry summers with temperatures that may exceed 40 °C at the Dead Sea shore area. The average amount of rainfall varies from 400 mm/year in the mountainous area to 100 mm/year at the Dead Sea shore line area [11,13].

Geologically, rock formations in Jordan are mostly sedimentary, belonging to the Paleozoic, Mesozoic, and Cenozoic ages during the phases of Paleotethys and Tethys transgressions and regressions [9,14]. In the studied areas sedimentary rocks are mainly Upper Cretaceous limestone, sandstone, and marl [9], covered in some areas by basaltic rocks that were formed during the Cenozoic. The fault systems in the study area have three main trends: (1) NW-SE which is parallel to the Red Sea and was created as a result of the Sea floor spreading; (2) a WNW-ESE fault system, which is perpendicular to the Dead Sea regional transform fault; and (3) a NNW-SSE fault system following the trend of the Dead Sea Transform DSR fault which is the most prominent fault system in the studied areas [9].

Topographically, landforms could be defined as natural physical features of the earth's surface that were generated as a result of the underlying geological structures and the climatic nature of the area, controlling the erosion of the different rock units [15]. Accordingly, Jordan has high relief topography because of movement along the DSR fault in addition to the nature of the rock units in the area [9]. The two cases studies therefore have high relief topography (Figure 2).

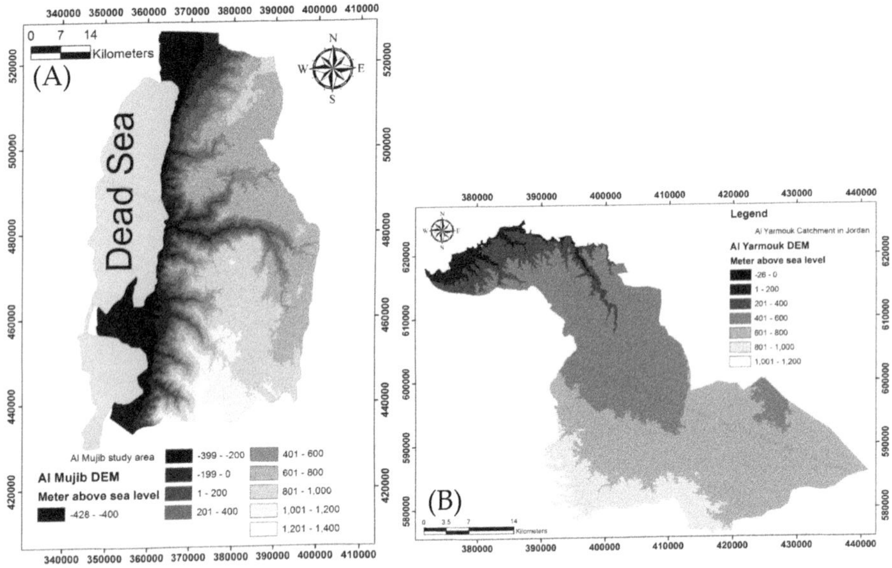

Figure 2. The high relief topography of the studied areas. (**A**) is the Mujib study area where the topography of the study area is effected mainly by the faults of the Dead Sea transform fault (a regional fault). (**B**) is the Yarmouk study area where the topography has effected mainly rock erosion, especially the sinkholes of limestone rocks.

3. Methodology

Geographic Information Systems (GIS) are computer-based systems for geo-referenced data. They are used for management (storage and retrieval), manipulation and analysis, and the display of geo-data [16]. Landscape Character Assessment (LCA) is a range of techniques used to classify, describe, and understand the evolution and physical and cultural characteristics of a landscape [17]. All the requested data to carry out a landscape character assessment must be geo-referenced in order to understand the spatial distribution of the various spatial characters [18]. Presenting the geo-data in the form of a vector has major advantages, such as having the original resolution without generalization and allowing efficient encoding of topology [19,20]. Therefore, we were trying to generate and have most of the geo-data that was requested in form of vectors (Shapfile by ArcGIS 10.2 software). The two cases studied have relatively large areas, about 4960 km^2, and complicated topographies. Moreover, there is a shortage in the high resolution of the land surface data. Therefore, a simplified LCA was done using GIS in order to generate LDUs [7,20].

Digital layers representing topography, geology, soil, land cover, and settlement patterns were used for creating LDUs at 1:250,000 scale. Contour lines at 50 m intervals were derived using ArcGIS software with Spatial Analyst extension (ESRI, 2013) from a digital elevation model available from the Advanced Spaceborne Thermal Emission and Reflection Radiometer (ASTER) at a 30 m spatial resolution. The contour lines were reclassified into elevation ranges to make the visual interpretation of landform easier. A slope raster was derived from the 50 m contour lines to provide more understanding about the steepness of the terrain by using Spatial Analyst extension. The slope raster was reclassified into percent slope ranges showing flat, undulating, and rolling slopes of land.

Digital geologic map layers were obtained from the Natural Resources Authority at 1:50,000 scale. The geologic map layers' attributes were too detailed for level 1 mapping; therefore, they were reclassified into larger rock units. Soil map layers at a scale of 1:250,000 were obtained from the national soil map and land use project implemented by the Ministry of Agriculture in 1994. Land cover data for the study area were obtained from the national land cover map of Jordan available at a scale of 1:250,000. Settlement Pattern data layers were created by manual digitizing of scanned topographic maps at a scale of 1:50,000. The settlement data layers available from topographic maps were updated by digitizing Google Earth imagery.

The mapping of LDUs was carried out by overlaying in GIS for the above-described layers and followed the MEDSCAPES LCA protocol [20]. This approach could be summarized in 4 steps as follows:

Step 1: Topographic analysis: the study area was divided into landform polygons representing similar physiography. Contour lines of Digital Elevation Model (DEM), of 30 m resolution, were used to identify similar landform units. A two digit code representing the landform was assigned to each polygon. Table 1 shows the landform types identified by the MEDSCAPES protocol, their description, and the number of units found of each class within the tested area.

Step 2: Ground type analysis: the landform polygons were subdivided according to the ground type. A code representing the ground type was assigned to each polygon in the attribute. The soil and geology layers were used to analyze ground type, and to decide whether the landform unit should be subdivided. Table 2 shows the ground types description and codes and the number of units found in each type.

Step 3: Land cover analysis: polygons resulting from the ground type analysis were again subdivided if land cover changed within each polygon. A code representing land cover was assigned to each polygon even if the units were not subdivided. Table 3 shows the land cover codes, the description, and the number of units found in each class.

Step 4: Settlement analysis: polygons resulting from land cover analysis were subdivided if the settlement pattern varied within each polygon. A field representing the settlement pattern was created, and the settlement pattern code was assigned to each polygon even if the polygon had a uniform settlement pattern and was not subdivided. Table 4 shows the settlement pattern codes, the description, and the number of units found in each class.

The two tested cases studies were divided accordingly into LDUs based on these four steps (Figure 3). However, the generated polygons that have an area less than the 5 km^2 were dissolved into the largest ones. This was in order to have a considerable area of homogeneous land description units. These units were attributed by the mentioned land surface geo-data. Afterword, the shape file of homogeneous landscape character was converted to a special geo-data format that is called Keyhole Markup Language (KML) format, which is compatible with Google Earth Google Earth and has a great capacity to calibrate and validate the landscape character units (because of its ability to zoom and observe the units from different directions, and to recognize the urbanized area and determine the land use [21]). Field work campaigns of two weeks were done in order to validate desk-top mapping. GPS equipment of 5 m accuracy was used to determine field sites and field work sheets were filled to describe each LDU in order to have a detailed description for each unit. This detailed description includes topography, land cover, soil, and geological units. The modifications and then corrections of LDUs were done with the help of Google Earth (Figure 4) and field work, saved in KML format and then imported into the ArcGIS 10.2, and re-edited where necessary.

Table 1. Landforms, descriptions, and the code of each class.

Landform	Description	Code
Lowlands	Areas of land generally lying less than 200 m above the drainage base-level. These can be further characterized as flat (<2% slope or 1:50), undulating to gently undulating (2–10%, 1:50–1:10), or rolling to strongly rolling(10–50%, 1:10–1:2)	Lf Lu Lr
Hills	Somewhat elevated tracts of land (around 200–600 m above) but also usually rolling slopes (10–50%)	H
High Hills	Somewhat elevated tracts of land (around 400–900 m above base-level) * but also steeply rolling to precipitous (>50%)	Hh
Uplands	Elevated and often extensive tracts of land with variable relief. Generally between 600–1000 m above base level	U
Mountainous area	Elevated tracts of land. Generally >1000 m and typically showing peaks, ridges, and steep or precipitous slopes	M
Plateau	The essential criteria for plateaus are low relative relief (extensive area of flat) and some altitude (e.g., over 200 m)	P
Eroded Plateau	Extensive areas of elevated land, undulating to rolling, often with dendritic drainage patterns	Pe
Valleys	An elongated hollow between hills or ridges. A stretch of a land watered by a river (at least seasonally) and enclosed or flanked by its erosion slopes	V
Escarpment	Extensive area at any altitude with average slope angles >50%. Often includes cliffs and rock outcrops	E

Note: * Each landform class description has a symbol of two letters in the geo-data base for labeling in the map.

Table 2. Ground type and the code of each class.

Soil Class	Code
Mixed Soils	M
Shallow Soils over chalk and limestone	R
Volcanic	V

Table 3. Land cover, description, and the code of each class.

Class	Description	Code
Artificial surfaces	Urban; industrial, commercial and transport units; mine, dump and construction sites; artificial non-agricultural vegetated areas.	A
Agriculture	Arable land, permanent crops, horticultural crops, , *enclosed* pastures, mixed agriculture.	G
Forest	Natural forests; Artificial forest (plantations).	F
Scrub	Includes maquis, garrigue and other mixed scrub. Typically sclerophyl shrub species attaining 1m+ in height.	S
Rangelands and rough grazing	Dwarf scrub, hedgehog scrub, phrygana often mixed with annual grasses and herbs , typically grazed by free-ranging goats and/or sheep. Vegetation usually <0.5 m in height., of aromatic, xerophytic and drought-adapted shrub species.	R
Bare land	Land with exposed bedrock, boulders, rocks, gravel etc *with virtually no vegetation* nor any specific clear use.	B
Wetlands& Water Bodies	Inland wetlands like marshes and peat bogs; coastal wetlands like salt marshes, salines and intertidal flats. Inland waters like water courses and water bodies; and marine waters like estuaries, and coastal lagoons.	W
Mixture of rangeland and bareland	Mixed landcover of rangeland and bare lands.	R-B
Mixture of rangeland and agriculture	Mixed landcover of rangeland and agriculture lands.	R-G

Table 4. Settlement pattern, description, and the code of each class.

Class	Description	Code
Unsettled	Areas with no settlements	U
Sparsely Settled	The typical distance between neighboring settlements is 5 km or more	Ss
Settled	The typical distance between neighboring settlements is less than 5 km	S
Urban	Cities, and other large built up areas greater than 10 km^2 in extent	Ur

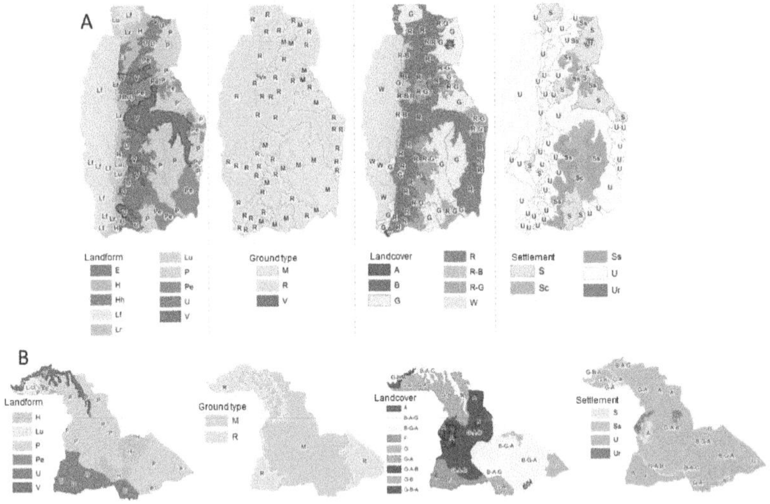

Figure 3. Land characterization based on landform, ground type, land cover, and settlement pattern. (**A**) is for the Wadi Mujib case study and (**B**) is for the Wadi Yarmouk case study. The meanings of letters are defined in the tables.

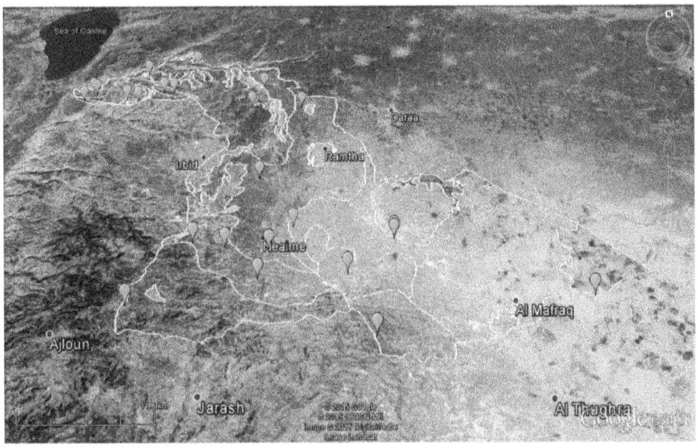

Figure 4. Google Earth image for the Yarmouk study area. The yellow line represents the border of LDUs and the green balloons represent the field work sites where we did the calibration for LDUs. The same method was used to calibrate the LDUs of the Wadi Mujib study area.

4. Results and Discussion

The study area of the Yarmouk River Basin was divided to 22 LDUs. Theses LDUs have an average area of about 63 Km^2, with a minimum area of c.5 Km^2 and a maximum area of c.372 Km^2. However, since the study is a comparative one, these LDUs were categorized in typological order, according to the maximum differences between the LDUs, to reduce the LDUs number and ease the comparison. Accordingly, the number of the LDUs in the Yarmouk study area become 14 typological units that are presented in Figure 5. However, the LDUs of the Wadi Mujib study area were 66. Theses LDUs have an average area of about 68 Km^2 with a minimum area of c.5 Km^2 and a maximum of c.634 Km^2. They were categorized to 14 typological too and are presented in Figure 6.

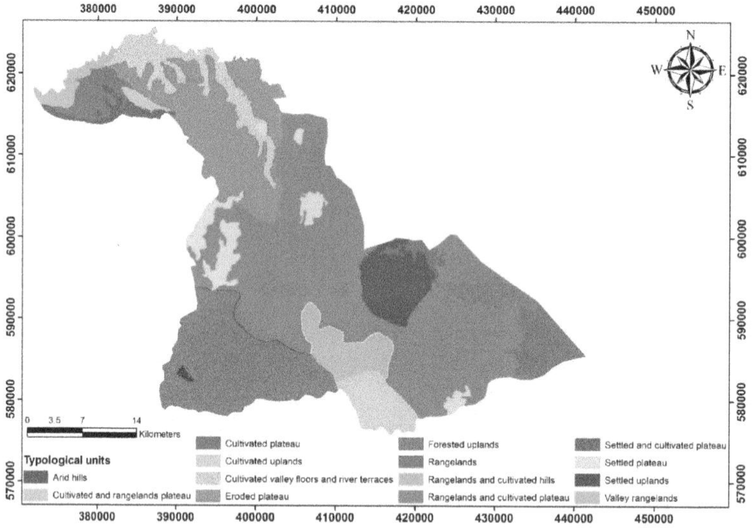

Figure 5. The 14 Typological units in the study area of Wadi Yarmouk. The numbers of units are increased in the north-west direction as a result of the topographical control.

In the Yarmouk River basin, the landscape can be broadly classified into three distinct regions: the southern and eastern highlands (forested uplands, settled uplands, rangelands and cultivated hills, and cultivated uplands and arid hills); the central plateau (rangelands and cultivated plateau, settled plateau, eroded plateau, and settled and cultivated plateau); and the valley area itself (cultivated valley floor and river terraces, and valley rangelands). LDUs follow the geology of the drainage basin, with the Middle Cretaceous limestones exposed in the centre of the dome of the Ajloun area in the south, becoming younger at the margins of the dome. Thus, the northern part of the drainage basin shows progressively younger exposures in the centre of the drainage basin (the Wadi Sir Formation) and in the north (the Amman Silicified Limestone Unit, the Um Rijam Formation, and the Eocene basalt flows) [9]. In the Mujib area, the landscape units can also be broadly divided into the upland plateau areas (settled plateaus, cultivated and mixed agriculture rangeland, arid high hills, eroded plateau, mixed farming and rangelands, semi-arid plateau rangelands, unsettled upland rangeland, semi-arid eroded plateau rangelands), the landscapes seen on descent into the Dead Sea (unsettled lowland, sparsely settled grazed and cultivated uplands, hill and valley rangelands, eroded plateau, mixed farming and rangelands), and the rift valley floor area.

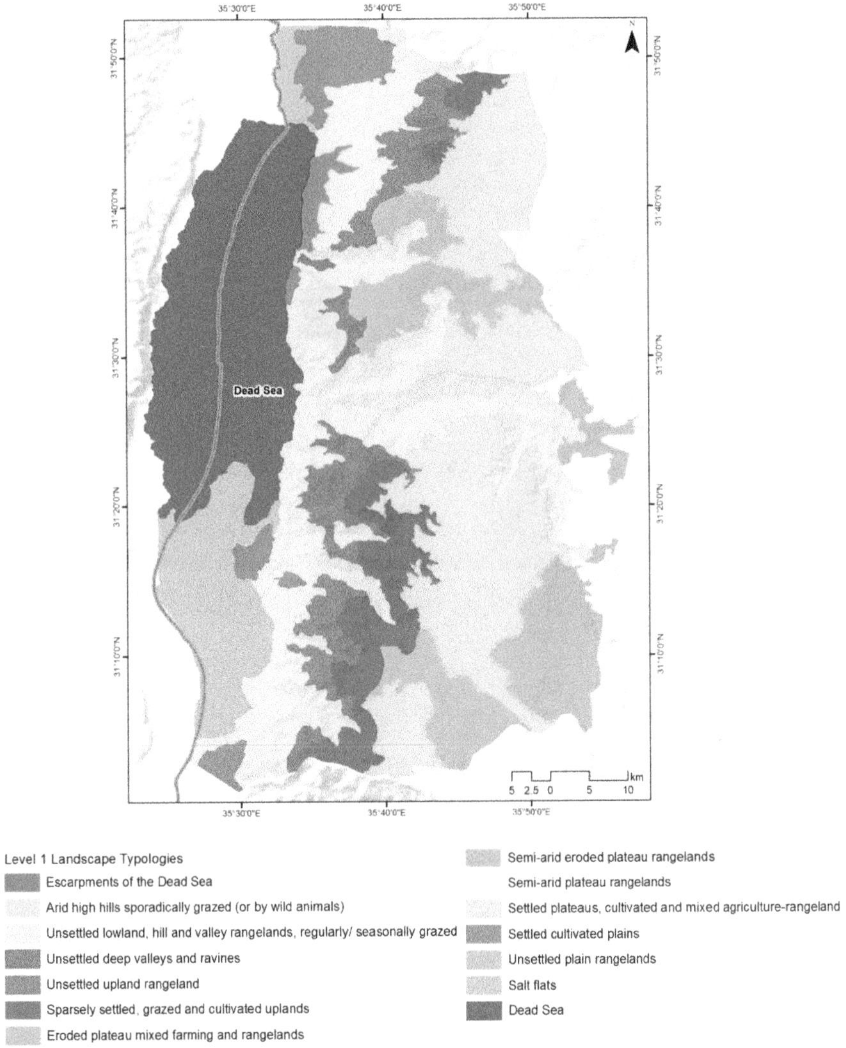

Figure 6. The 14 Typological units in the study area of Wadi Mujib. The numbers of units are well distributed in all the directions in this study area.

There are several factors that control the type, area, and distributions of the landscape units such as climatic conditions, land cover, land use, soil types, and outcropped rocks units [18,22]. All of these factors are controlled by topography, especially when there is a high relief topography such as in the two case studies that we studied. For instance, the elevations in the topographical units control the climatic parameter of temperature degrees and the rainfall yearly quantity [23]. Plant cover types and distribution are affected by rainfall regimes and temperature [24,25]. Furthermore, landforms that are generated by topography are one the main factors that control the soil texture types, as well as the occurrence of outcropping rocks [18,26].

The Wadi Mujib study area is located within a regional tectonic area that has conservative plate movement through the Dead Sea Rift [9]. These tectonic movements created different local faults within the Mujib study area. These faults are the main reasons for the high relief topography of the Wadi Al Mujib study area. The elevations of the Wadi Mujib study area vary from 420 m below sea level to about 1400 m above sea level [9,12,27]. The effect of the Dead Sea transform fault in the Yarmouk study area is less than the Al Mujib study area because it extends farther away from it [28]. Therefore, the Yarmouk area has lower relief topography than the Wadi Mujib study area. The elevation of the Yarmouk area varies from 26 below the sea level to 1200 m above the sea level.

Faults as an underline structure are one reason that generates the topographical units of any landscape [28]. However, the local faults (strike slip faults, mainly) of the Wadi Mujib study area bonded the plateau landform in several sites such as Wadi Mujib and Wadi Zerka Ma'in [12]. However, erosion occurs through these faults during the winter by overland flow and flash floods [11,24]. Hence, the largest landscape typological unit in the Wadi Mujib study area is the eroded plateau. However, comparable to the Wadi Al Yarmouk study area, this landscape typological unit is limited since the Wadi Yarmouk study area has fewer faults as a result of not being within the Dead Sea transform fault [27].

The largest landscape typological unit in the Yarmouk study area is the rangeland and the cultivated plateau. The plateau was originally a flat topographical unit [28]. However, the flat area has a horizontal slope, and this would reduce soil erosion; hence, this unit has a considerable thickness of soil of about 6 m, more than the other landscape typological units, and is enriched by agricultural cultivation and rangeland use [18].

The Yarmouk study area is nearer to the Mediterranean Sea, and therefore this study area is more affected by the Mediterranean climatic conditions than the Wadi Mujib study area, which is characterized by arid and semi-arid climatic conditions, mostly [11,13]. Therefore, the Wadi Yarmouk study area receives rainfall quantity more than the Wadi Mujib study area [27]. However, the uplands in the Yarmouk study area has ecosystems in the form of forested upland, which is common for upland in the Mediterranean climatic zone, while the upland and the Plateau area in the Wadi Mujib study area are mostly range lands, which common for arid and semi-arid regions [23].

The water resources are mainly in the form of groundwater in both of the cases studies [13,27]. However, the groundwater level in the Wadi Yarmouk study area is higher than the groundwater level of the Wadi Mujib study area because of the higher groundwater recharge availability during the winter season [29]. The suitable Mediterranean climatic conditions, and water resource availability have caused the Wadi Yarmouk study area to become settled and urbanized more than the Wadi Mujib study area.

In the Wadi Mujib study area, the rapid decrease of the Dead Sea level (1 m per year) during the last 50 years generated the Dead Sea escarpment typological unit that has sinkholes as a result of that decrease, while the mining of potash generated the salt flat typological unit. These two typological landscape units are associated with the Dead Sea [30].

Spatial planning in these and other areas of Jordan has tended to ignore or underestimate the importance of geological hazards. This may be understandable, given that geologists are typically not involved in the planning process. However, the adoption LCA as a planning tool is an important opportunity to avert some of the problems that have occurred in the past. In looking at the LCA maps produced for the Yarmouk and Mujib areas, the various lowland designations are within or near the bottom of the Dead Sea Rift system or one of the tributaries. The valley floor sediments in these areas tend to be unconsolidated alluvium, which suggests that these areas, while suitable for agriculture, could be problematic for building due to problems such as salinity, difficult foundation design, and seismic amplification. In some areas, particularly near the margins, flooding may be an issue as well. Therefore, the designations of valley rangelands, cultivated valley floors, and river terraces are rangelands that need to be marked for these issues in the Yarmouk area. Similarly, the salt flats, the settled cultivated plains, and unsettled lowlands in the Mujib area should be watched for the same issues. In addition, the salt flats have serious problems with subsidence associated with the fall of the Dead Sea water level [31].

Hills, high hills, and uplands pose different issues. In particular, relief in these areas results in issues such as soil loss. The arid hills and the forested uplands of the Yarmouk area fall within this physiographic area. The forested uplands here are a major recharge area for the groundwater through the entire basin [29], because of the high levels of rainfall there. The arid high hills of the Mujib and Yarmouk areas also fall within this category, and tend to be over-grazed due to the lack of interest in agriculture or soil conservation in these areas.

The plateau and eroded plateau areas are also important recharge areas [32], occurring mainly in the Yarmouk basin. Thus, planning should take into consideration groundwater resource maintenance. Moreover, in areas where the bedrock is limestone such the Yarmouk basin, karstification can be an issue. Although karst provides for the availability of good aquifers, these are susceptible to pollution because contaminants move easily within them. Moreover, sinkholes, a typical karst feature, may present an obstacle for land-use planning. For example, in the city of Karak, an entire new neighbourhood known as Marj has grown to the east of the city. This neighbourhood suffers from unstable foundations and many buildings have had to be abandoned because they have been deemed unsafe. In addition, these tend to be areas where there are thick soils that are used for dryland agriculture. On the other hand, these areas tend to host major cities in the two areas (Irbid and Ramtha in Yarouk and Madaba in Mujib). Thus, urban expansion has led to the loss of prime agricultural land and the loss of recharge zones in both areas, creating the settled plateau areas in the Yarmouk and Mujib basins.

5. Conclusion and Recommendations

Landscape types demonstrate the strong influence of the topography, which is generated mainly by the underlying geological structures. This control is reflected in the spatial distributions of landforms, climatic zones, water resources, and plant cover. An optimized spatial plan must have a landscape character assessment and a landscape units map in order to investigate the optimized sites for water, environmental, and industrial projects. Moreover, landscape types play a crucial role in increasing understanding of the distributions, and therefore in supporting biodiversity. And should be also taken into consideration when planning for nature conservation.

In Jordan, we found that the Wadi Mujib study area must have more projects that carry out a sustainable water resources management, while in the Wadi Yarmouk study area we found that it must be a land policy that stops the occupation of the agricultural land by the urbanized area.

Based on the application of LCA, the following can be recommended for improved spatial planning in the two tested areas:

5.1. In the Wadi Mujib study area:

(1) There must be a land policy that restricts any further development in the landscape type of the Dead Sea escarpment, since it is a unit of sinkholes where constructions run a subsidence risk.

(2) There must be a strategy that encourages artificial groundwater recharge and water harvesting on the eroded plateau unit to improve the water resources and agricultural activities there.

5.2. In the Wadi Yarmouk study area:

(1) There must be a land policy that controls urban expansion from extending onto the settled plateau landscape unit and the cultivated and rangelands plateau unit. Otherwise, the area of the agricultural activity would decrease considerably. However, this would also decrease the groundwater recharge, since the runoff would be increased.

(2) The forested upland unit must be considered as a conservation area and protected from any construction or industrial activity. This unit is rarely found in Jordan and includes different ecosystems.

The study demonstrated, in both cases, that groundwater protection, slope stability, sinkhole avoidance, flood control, and the protection of agricultural land would highly benefit from the insight generated by the LDUs and the LCA process and, in particular, the role of geology. Further research must be carried out in order to evaluate the planning improvement sections in Jordan precisely, and to evaluate other possible factors in addition to geology.

Acknowledgments: This study was part of the MEDSCAPES project funded by the ENPI-CBCMED programme. We are thankful to the former director of Environmental Study Centre of Terra Cypria Mr. Nicholas Symons for his help to do this research. We thank two anonymous reviewers whose comments and suggestions helped to improve the manuscript.

Author Contributions: All authors contributed and approved final manuscript before submission.

Conflicts of Interest: The authors declare no conflict of interest.

References

1. Abu-Jaber, N. *Jordan and the Environmental Challenge (الاردن و التحدي البيئي)*; Dar Al Shorouk publishers, Amman, with support from the Ministry of Culture: Amman, Jordan, 2010; p. 216. (In Arabic)
2. Potter, R.B.; Darmame, K.; Barham, N.; Nortcliff, S. "Ever-growing Amman", Jordan: Urban expansion, social polarisation and contemporary urban planning issues. *Habitat Int.* **2009**, *33*, 81–92. [CrossRef]
3. Al-Eisawi, D. *Vegetation of Jordan*; UNESCO- Regional Office for Science and Technology for the Arab States: Cairo, Egypt, 1996.
4. Beauregard, R.A.; Marpillero-Colomina, A. More than a master plan: Amman 2025. *Cities* **2011**, *28*, 62–69. [CrossRef]
5. Department of Statistics. *Statistical Yearbook*; Government publication: Amman, Jordan, 2014; Chapter 5; p. 200.
6. Tarabieh, S. Sustainable Spatial Planning Constraints in Al-me'raad Watershed, Jordan. Unpublished M.Sc. Thesis, German Jordanian University, Amman, Jordan, 2015.
7. Abu-Jaber, N.; Abunnasr, Y.; Yahya, A.A.; Boulad, N.; Christou, O.; Dimitropoulos, G.; Gkoltsiou, K.; Khreis, N.; Manolaki, P.; Michael, K.; et al. Travelling in the eastern Mediterranean with landscape character assessment. In Proceedings of the Third International Conference on Remote Sensing and Geoinformation of the Environment, Paphos, Cyprus, 16–19 March 2015.
8. Perdew, L. *Understanding Jordan Today*; Mitchell Lane Publishers, Inc.: Newark, DE, USA, 2014; pp. 17–18.
9. Bender, F. *Geology of Jordan. Contribution of the Regional Geology of the Earth*; Borntraeger: Berlin, Germany, 1974; p. 196.
10. Moshrik, H.R.; Abu-Allaban, M.; Al-Shayeb, A. Climate change in Jordan: A comprehensive examination approach. *Am. J. Environ. Sci.* **2009**, *5*, 58–68.
11. Odeh, T.; Rödiger, T.; Geyer, S.; Schirmer, M. Hydrological modelling of a heterogeneous catchment using an integrated approach of remote sensing, a geographic information system and hydrologic response units: The case study of Wadi Zerka Ma'in catchment area, north east of the Dead Sea. *Environ. Earth Sci.* **2014**, *73*, 3309–3326. [CrossRef]
12. Odeh, T.; Gloaguen, R.; Schirmer, M.; Geyer, S.; Rödiger, T.; Siebert, C. Investigation of catchment areas migrations through a sinstral and dextral strike slip faults: The case study of Zerka Ma'in and Al Hasa catchment areas, East of the Dead Sea in Jordan. In Proceedings of the 4 ASME/WSEAS International conference on geology and Seismology (GES'10), Cambridge, UK, 23–25 February 2010; pp. 120–126.
13. Salameh, E.; Bannayan, H. *Water Resources of Jordan—Present Status and Future Potentials*; Friedrich Ebert Stiftung: Amman, Jordan, 1993.
14. Abu-Jaber, N.S.; Kimberley, M.M.; Cavaroc, V.V. Mesozoic-Palaeogene basin development within the Eastern Mediterranean borderland. *J. Petrol. Geol.* **1989**, *4*, 419–436. [CrossRef]
15. Dunne, T. *Evaluation of Erosion Conditions and Trends*; FAO Conservation Guide: Rome, Italy, 1977; pp. 53–83.
16. Kam, T. Integrating GIS and remote sensing techniques for urban land-cover and landuse analysis. *Geocarto Int.* **1995**, *10*, 39–49.
17. Swanwick, C.; Land Use Consultants 2002. *Landscape Character Assessment: Guidance for England and Scotland*; The Countryside Agency/Edinburgh; Scottish Natural Heritage: Cheltenham, UK, 2002.
18. Zube, E.; Sell, J.; Taylor, J. Landscape perception: Research, application and theory. *Landsc. Plan.* **1982**, *9*, 1–33. [CrossRef]
19. Chou, Y. *Exploring Spatial Analysis in GIS*; Onword Press: New York, NY, USA, 1997; p. 500.

20. Brabyn, K. Solutions for characterizing natural landscapes using Geographical Information Systems. *J. Environ. Manag.* **2005**, *76*, 23–34. [CrossRef] [PubMed]
21. Dodsworth, E.; Nicholson, A. Academic Uses of Google Earth and Google Maps in a Library Setting. *Inf. Technol. Libr.* **2012**, *31*, 32–36. [CrossRef]
22. Giles, R.; Trani, M. Key elements of landscape pattern measures. *Environ. Manag.* **1999**, *23*, 477–481.
23. Kepner, W.; Watts, C.; Edmonds, C. A landscape approach for detecting and evaluating change in a semi-arid environment. *Environ. Monit. Assess.* **2002**, *64*, 179–195. [CrossRef]
24. Gustafson, E. Quantifying landscape spatial pattern: What is the state-of-the-art? *Ecosystems* **1998**, *1*, 143–156. [CrossRef]
25. Ferguson, B. Landscape hydrology. *J. Environ. Syst.* **1991**, *21*, 193–206. [CrossRef]
26. Lausch, A.; Herzog, F. Applicability of landscape metrics for the monitoring of landscape change: Issues of scale, resolution and interpretability. *Ecol. Indic.* **2002**, *2*, 3–15. [CrossRef]
27. Horowitz, A. *The Jordan Rift Valley*; Ballkema: Rotterdam, The Netherlands, 2001; p. 730.
28. Keller, E.A.; Pinter, N. *Active Tectonics: Earthquakes and Landscape*; Prentice-Hall: Upper Saddle River, NJ, USA, 2002; p. 362.
29. Abu-Jaber, N.; Kharabsheh, A. Ground water origin and movement in the upper Yarmouk Basin, Northern Jordan. *Environ. Geol.* **2008**, *54*, 1355–1365. [CrossRef]
30. Niemi, T.; Ben-Avraham, Z. *The Dead Sea: The Lake and Its Setting Monographs on Geology 442 and Geophysics Press 36*; Oxford University: New York, NY, USA, 1997; pp. 326–336.
31. Closson, D.; Karaki, N.A.; Klinger, Y.; Hussein, M.J. Subsidence and sinkhole hazard assessment in the southern Dead Sea area, Jordan. *Pure Appl. Geophys.* **2005**, *162*, 221–248. [CrossRef]
32. Al Qudah, K.; Abu-Jaber, N.; Jaradat, R.; Awawdeh, M. Artificial rainfall tests, soil moisture profiles and geoelectrical investigations for the estimation of recharge rates in a semi-arid area (Jordanian Yarmouk River Basin). *Environ. Earth Sci.* **2015**, *73*, 6677–6689. [CrossRef]

land

MDPI

Article

A Stakeholders' Analysis of Eastern Mediterranean Landscapes: Contextualities, Commonalities and Concerns

Theano S. Terkenli * and Dimitris Kavroudakis

Department of Geography, University of the Aegean, 81100 Mytilini Lesvos, Greece; dimitrisk@aegean.gr
* Correspondence: t.terkenli@aegean.gr; Tel.: +30-225-103-6414

Received: 6 November 2017; Accepted: 8 December 2017; Published: 14 December 2017

Abstract: This study aims at demonstrating and critically assessing high-level landscape stakeholders' perceptions and understandings of landscape-related issues, threats and problems, in the Eastern Mediterranean, through a purposive comparative research survey of four case studies: Cyprus, Greece, Jordan and Lebanon. Employing qualitative data analysis of intensive stakeholder interviews, performed in the broader context of the MEDSCAPES ENPI-MED project (www.enpi-medscapes.org), the paper draws together the insights and concerns of a total of 61 public entities, private entrepreneurs, academicians and NGO representatives, on landscape knowledge, understanding, management and public awareness, in these four countries. The results point to significant commonalities among them and begin to show relational and synthetic nature of the interrelationship between humans and the landscape, as it developed in the context of the local and regional geographies and histories of this broader region, affected by and involving a series of relevant geophysical, economic, political, social, moral, institutional and other parameters.

Keywords: landscape; stakeholders' analysis; Eastern Mediterranean; Greek-speaking; Arabic-speaking

1. Introduction and Context of the Study

Having been established as a basic good in the western world, landscape has infiltrated as a concept, a construct and an area of spatial analysis and intervention, elsewhere, in various cultural realms and non-western worlds. The propinquity of the Middle East to Europe and the longstanding interchange between Arab-speaking and Greek-speaking peoples and ways of life and thought has inadvertently influenced landscape notions, values and practices.

With the aid of a stakeholder survey, this paper sets out to assess precisely those commonalities and differences between these two worlds, as regards the landscape concept and management/planning, as the western notions of landscape increasingly take hold also in this part of the world. For this purpose, the paper presents and critically discusses high-level landscape stakeholders' knowledge, perceptions and understandings of landscape-related issues and problems, in the Eastern Mediterranean, through a purposive comparative research survey in four countries (Cyprus, Greece, Jordan and Lebanon). The study, thus, begins to demonstrate the deeply relational character of the constitution of people's understanding of and interrelating with the landscape on a broader physical and cultural geographical basis.

The European Landscape Convention [1] has been widely and rightly acknowledged—at least in Europe—as our most useful and comprehensive tool in setting out the (Western) principles for landscape conceptualization, valuation, planning, use and management. The ELC's successful negotiation of the (Western) landscape, generally speaking, has inadvertently further entrenched the hegemony of the Western concept and cultural construct of Landscape, displacing or marginalizing local and vernacular landscape notions and terms, around the world. Despite growing acknowledgment of such processes [2–4], a great deal of these indigenous, organic ideas and uses of landscape-related terms still goes unnoticed and unaccounted for [5]. Arguing for situated, contextualized landscape ideas, within their geographical, environmental, cultural and political framework and context, we uphold the most widely-acknowledged premises of landscape definition. These premises rest on the symbiotic relationship of humans with their physical surroundings that is the physical/natural and human/cultural interconnections and interactions, co-existing and co-evolving through time (relational and holistic nature of Landscape). The absence of an Arabic word for 'landscape,' for instance, as this paper upholds, is most likely to be due to cultural differences between a Middle-Eastern conception of 'landscape' and a Western one. Moreover, the case of the Greek-speaking world appears to be especially complex, due to variable and multiple historical and geographical influences imbued on the processes of formulation of a landscape concept/notion/conscience—as concerns both the actual condition of the Greek landscape and its references and representations—as compared to other (western or northern) such European experiences.

This study was undertaken within the context of an ENPI-MED EU-funded research project (MEDSCAPES www.enpi-medscapes.org). At the outset of this larger project, a questionnaire survey of high-level landscape stakeholders was undertaken in the four partner countries (Greece, Cyprus, Lebanon, Jordan), aiming at, among other things, unveiling high-level landscape-related stakeholder perceptions and understandings of landscape-related issues, threats and potential, in their countries. For this purpose, after a brief theoretical investigation of the Arab and Greek 'landscape' concepts, the article proceeds to the analysis and discussion of its survey findings, with the aid of empirical data and hermeneutical tools.

2. The Western Idea of Landscape

Landscape, a Western—and specifically European—idea, evolved over time, shaping and reflecting the evolution of the European landscape itself [6,7]. One of the oldest and most succinct definitions of landscape in the Western world, «a portion of land which the eye can comprehend at a glance» [8], places great significance to landscape as a visible expression of the humanized environment perceived mainly through sensory and particularly visual, as well as cognitive processes. The visual landscape dimension refers to landscape form—or else, structure—and its 'objective,' tangible entity—to distinguish it from its intangible, perceptual and symbolic dimensions, also addressed in the ELC.

Furthermore, landscapes are created by human action and experience inscribed in place through time, enhancing and upholding human livelihoods. Thus, our ways of conceiving the tangible landscape (or landscape dimensions) may broadly be summarized as (a) visual/perceptive and (b) performative/experiential (interactive or relational nature of landscape). In reality, since both of the latter dimensions involve human senses and processes of cognition, they interweave in complex and multi-relational ways of landscape conceptualization, intricately relating the subject (observer, user or visitor) with the object of perception, intervention or pleasure (landscape) [9–12]. We may, thus, talk about multiple 'landscape spatialities,' the term Cosgrove employs in his discussion of the succession of landscape types in European land use and spatial design history [13]. Accordingly, landscape in the Western world has not been simply conceived as a piece of land (territory) or its appearance (view, prospect or scenery) [12]. It has also been defined in terms of "a nexus of community, justice, nature and environmental equity, a contested territory ... " [7], raising issues of shared values, belonging and identity [14], while bearing the imprint of power and control of its dominant governing or elite classes [15].

To sum up, landscape is not just the concept of the environment, as modified by humans [16]; according to the ELC, it also encompasses an intangible, (inter)subjective perspective, people's perception of it [17]. Further, it is a real, contingent, embodied, reflexive and unfolding entity. In recent geographical discourse, newly-emerging more-than-representational and enacted landscape geographies highlight the relational constitution of the landscape, signaling a return to its more organic, bottom-up experiential nature (the humanistic tradition). Rather than landscape beheld, what is advocated in these new post-phenomenological landscape geographies is landscape inhabited, engaged with and processed [18]. Thus, contemporary geographic thought, reflecting broader scientific trends, moves beyond naturalistic views of landscapes as neutral, external backdrops to human activities and culturalistic views of landscapes as cognitive/symbolic orderings of space and advocates an enacted or 'dwelling perspective' of landscape constitution [19].

The Western way of looking at a landscape from a geographically fixed point (a 'view point' offering a 'perfect view'), historically laden with specific aesthetic and geometrical principles, is not universal. The ideas encapsulated in the single word 'landscape' may, in other cultural systems, be communicated with the aid of a variety of other terms or meanings, while, at the same time, these other cultures may employ landscape notions and meanings absent from the Western term 'landscape.' Well-documented cases, where Western notions of landscape were transposed in other cultural systems, indicate a differential historical evolution, as regards the adoption of such notions, often accompanied by linguistic losses or by processes of linguistic assimilation or acculturation, usually leading to the impoverishment of indigenous alternative landscape notions and practices [5,20]. For instance, the Western concept of landscape, always implying some sort of geographical delimitations, differs from South Asian and Chinese concepts that include, among others, undetermined entities without visible borders [5], territories, or people's adaptation to specific (natural) local environments [21]. Similarly, Bruns and van den Brink discuss the inherent absence of the 'landscape' concept in the languages of the larger Near East, including Farsi and Turkish. In Turkey, for example, the Western term 'Peyzaj' was introduced in the 1920s and evolved to replace indigenous landscape-related terms of Arabic origin. In the process, however, its use gradually encapsulated local conceptions of 'garden' and 'land,' long-standing societal priorities for a society that eventually 'owned' the Western term, in a way adjusting it to its own cultural system [5].

3. Landscape in the Middle East—The Arab Concept

There is no equivalent word to the English term "landscape", in the Arabic language. In our discussion of the ways that landscape notions have been conceptualized and used in the Middle East and specifically in Jordan and Lebanon, we will rely on the outstanding groundwork laid out by Jala Makhzoumi, an influential landscape researcher and scholar, in this area of scientific inquiry. Makhzoumi recounts that available translations are generally restricted and outdated, as they essentially adopt the 18th-century English, scenic meaning. The Arabic translation of "landscape", then, has been variously offered as "natural scenery" [20], "land scenery" (Baalbaki & Baalbaki 1997 and Nasr & Khatib 1985, both in the work of Makhzoumi [20]) and "view of the countryside" [20]. In Arabic culture, however, landscape is not simply a distant viewing experience: "[Landscape] is appreciated bit by bit, through movement in space and time and an engagement of all senses" [20] (p. 222), [22]. In place of the seeming absence of a corresponding Arabic term, an array of terms has been traditionally used by Middle Eastern peoples, such as hada'ik (gardens), fadha'at kharijia (outdoor spaces) and bi'a (environment), connoting one or more meanings of landscape [20] (p. 223). For example, according to the same author, the word "jenna" in Arabic, is used for paradise and it is also a word used for garden; it is once a physical place and a conceptual space, a state of peace and contentment.

Thus, alleged "shortcomings of the Arabic translations can be attributed in part to the complexity of the English word, which makes it impossible to combine the different meanings into a single translation. However, the main shortcoming is due to the difficulties inherent in transferring a complex concept, such as 'landscape', from one socio-cultural context to another ... intentionally or inadvertently [superimposing] ideas and conceptions ... alien to the context of the Middle East. An extreme case in point is Orientalism, a European invention that served as a way of coming to terms with the Orient" [20] (p. 218). Such a reading of the Middle Eastern landscape was used not only to dominate a place and culture [23] but also, as argued by Said, to displace "Oriental history as a history possessing its own coherence, identity and sense" (1985: 86).

Indeed, given the colonial history of this part of the world, Makhzoumi [24] argues for a dual understanding, conceptualization and practice as concerns the landscape, geographically distinguished and anchored in the rural-urban dichotomy of the broader region. The uneasiness and tension in the employment of the different terms implying 'landscape' presented above are the outcome of the strenuous coexistence of two conceptions of landscape in the Arab Middle East: "The first, a local interpretation of the English word, dates back to colonial and post-colonial rule, in the first half of the twentieth century ... borrowed from Western culture with total disregard to the social and environmental context ... [as opposed to] another, less recognized conception that is rooted in long-standing perceptions and vernacular valuations that evolved in the Middle East over millennia. Both conceptions struggle to find [their] correct linguistic expression. The borrowed conception searches for a word in Arabic that encompasses the layered meaning associated with the English word ... with little success. Neither is there an all-encompassing word in Arabic for the 'rooted conception'" [24] (p. 112).

The landscape conceptualization "borrowed" from the Western definition of landscape has taken root in the colonized, westernized part of the Arabic world (mainly the cities but also all territory where Western planning, design and management rules and norms apply, such as parks, lawns, open spaces, etc.), whereas the "rooted" landscape conceptualization still reigns in environments where "traditional" (but highly dynamic) pre-colonial ways of life persist (mainly the rural world) [24]. The former landscape conceptualization privileges the more distant, visual/perspectival and aesthetic 'way of seeing' the world [25], which predominate in Western ways of viewing the landscape (as illustrated in the ELC landscape definition), whereas the more organic latter conceptualization of landscape privileges the embodied, reflexive, integrative and instrumental ways of living in and experiencing/relating with it, with all the senses. The former originates in a view from above, or better an oblique birds-eye-view opening up the landscape to its viewer [26], whereas the latter stems from a view from within.

As described above, the Middle Eastern concept of "landscape", on the contrary, allows for "an experiential appreciation of one's surroundings, bit by bit, through movement in space and time, as well as through engagement of all the senses". As such, it cannot be perceived in one "glance", nor can it be appreciated through the Westerner's distant gaze.

Supporting Makhzoumi's arguments, we base our study design and analysis on the fact that the 'traditional' reciprocal relationship between society and space and its concomitant landscape interpretations were severed, after years of colonial domination and recent integration into global systems, as socio-economic and political changes in the contemporary Middle East transformed culture and landscape alike [20] (p. 219) and alienated the peoples of the colonized territories from their own places. Twentieth-century population increase is changing the rich and extremely heterogeneous physical regional landscape [20] (p. 219), by replacing traditional, vernacular patterns, both rural and urban, with homogeneous, contemporary ones [27]. Furthermore, the struggle against aridity has, over the centuries, influenced socio-political organization and shaped cultural expressions and values, including mythological narratives of the broader region. For instance, the design of urban landscapes and gardens as "earthly paradises" [28] in Arab-Islamic culture has been guided by the dictates of aridity, also amply expressed in Arab-Islamic art.

In the hostile environment of the Middle East, comfort and security and, by association, beauty, are embodied by landscapes that are human-modified and human-made—the colonial agrarian landscape and urban one respectively, in which nature has been "tamed", enclosed and ordered. We contend, that the "outward movement" of perceiving "landscape" as a scene that reaches its end in the horizon, is inevitably "introverted" in the Middle East—as are pictorial landscape representations depicted in Greek Orthodox ecclesiastical art, heavily indebted to Middle-Eastern aesthetic norms and practices. Such spatial and landscape interpretations and representations that privilege the spiritual realm and aim to convey an embodied sense of the divine. Nor are "horizon" or perspective as important landscape features, as they are in Western culture, where they imply human control and supervision (surveillance) of the land. Conclusively, according to Makhzoumi [21] (p. 222), the aesthetic meaning of "landscape" in the Middle East is fundamentally different—physically, perceptually and symbolically—from the Western scenic, "extrovert" historical and contemporary meaning of the word, also permeating contemporary Greek culture. We turn to the latter, next.

4. Landscape in SE Europe—The Greek World

The contemporary urban societies of post-war Greece and Cyprus adopted Western landscape norms and values, in a top-down approach, as these evolved in Europe since the Renaissance and are eloquently laid out in the ELC [1]. Although both Greece and Cyprus have signed and ratified the ELC, landscape planning and management is not high on the political agenda of either one of the two countries, as in many other Mediterranean countries [29] (p. 290). Despite the fact that the term "landscape" is now increasingly or implicitly mentioned in various designated protected areas and policies, in both countries, little effort is to put into managing it in an integrated and sustainable way, according to European-level norms and practices [30,31]. This may be due to either low prioritization and to a deficient lay landscape conscience, or to the lack of funds and concrete/concerted approaches to landscape management [32,33].

In the case of Greece, although a Landscape Committee was formed in the then Ministry of the Environment, Energy and Climate Change, for the first time in 2010, it barely begun to operate towards the implementation of the ELC, before it became ineffective, due to the country's plunge into "the crisis". In Cyprus, the authority responsible for the implementation of the ELC is the Department of Town Planning and Housing of the Ministry of the Interior, whose main scope of operation concerns urban and spatial planning. Recent attempts from Town Planning have resulted in the first landscape typology produced for the whole of the island, with a view to using it as a tool for Territorial Planning; the second part of the project on landscape typology was carried out during the MEDSCAPES project, while—despite a genuine interest from the competent authority—progress is slow in integrating landscape assessment in planning policy and legislation [31,34]. A further obstacle, in the case of Cyprus, is that landscape continues to mean different things to different disciplines or stakeholder groups, as demonstrated by Vogiatzakis et al. [31], or to be narrowly defined mostly in terms of its physical attributes but also highly tinged with identity issues and memories of the island's occupation and displacement. What follows is a more detailed account of the case of Greece; the case of Cyprus resembles the one of Greece, although Cyprus is one step ahead of Greece in terms of ELC implementation—at least as far national-level landscape mapping and Landscape Character Assessment are concerned.

A first but especially revealing view of the Greek landscape, as concerns its symbolic and representational properties and perspectives, emerges through landscape painting at the time of the formation of the new modern nation-state of Greece, after its war of independence in the 19th century [35] (p. 128). If landscape is a "way of seeing" closely connected to the development in Europe of modern urban, commercial life, then landscape representations in art are renditions in some form of images of the ideological construction of the newly-emergent European nation-states [6]. Analysis of 19th century Greek landscape painting by Western artists exposes the construction and depiction of the contemporary Greek landscape as a context of human life and experience in accordance with romantic ideals. At the basis of the emergent Greek (and Cypriot, for that matter) cultural identity were two ideals: (a) classicism; and (b) orientalism [36], expressing contemporary Western views and ideals of the modern Greek state and identity [37,38].

In the 20th century, Greek landscape painting gained only partial and gradual emancipation from the influences of foreign schools, through the development of various indigenous forms of expression [39]. According to Terkenli [35], however, the landscape ideal and form of representation most influential upon the Greek psyche and most characteristic of the Greek cultural realm remained the two-dimensional, apparently flat but actually inverted, perspective of Greek Orthodox art. The human figure tends to dominate in ecclesiastical iconography, rather than the landscape per se. This inverted and introspected perspective pulls the viewer into the painting, as mentioned in the previous section, "placing" the viewer at the center of the work of art and thus exerting a great power of suggestion over him or her. The creation of the Modern Greek state signaled not only the importing of foreign painting but also all manner of relevant spatial intervention to the Greek landscape ('top-down' western influences), as also happened in Cyprus. Nonetheless, Greek Orthodox art, surviving and flourishing amidst the deeply religious populace under Islamic domination, seems to have imprinted its highly influential worldview on the Greek mind and psyche ("bottom-up" concept and construct), providing an ideal and a way of relating to the world still pervasive in Greek life and art [35].

Although any sort of spatial conscience generally attributed to a cultural system tends to find its roots in the history of a modern nation-state, caution must be exercised in generalizing and totalizing as regards whole cultures or social systems. Before the post-war era of rapid Greek urbanization, Greek people of rural, mountain, or island pre-industrialized communities tended to live under conditions tightly interwoven and in harmony with their particular environments and landscapes [32,40], as was the case throughout pre-industrial Mediterranean Europe [34,41]. These relationships were inevitably upset or significantly altered, not through colonization but through occupation.

Under Ottoman occupation and cultural stagnation from the mid-15th to the early or mid-19th century, Greece never went through any of the stages of landscape planning and management and a development of landscape conscience that most modern Northern and Western European cultural landscapes had gone through by the 17th century—notions that accompanied the development of the latter landscapes up to our times [6,42]. Rather, Greece adopted from the West aspects of modernity in certain realms of life a posteriori, by implanting and overlaying them on to pre-existing cultural particularities and local ways of life. Despite the fact that the word "topio" (signifying landscape in the Modern Greek language) translates into a miniaturization of place, it nonetheless carries all the predominantly visual, aestheticized/scenic and distanced connotations of the Western landscape concept [13].

This landscape conceptualization has predominated in post-war rapidly-urbanizing Greece, gradually gaining a stronghold in the Greek society, in lieu of a series of environment-, locality- and place-related terms, which used to be employed by pre-industrial, "traditional"-though, certainly, not static-Greek societies in reference to the landscape. The "new", naturalized as the norm, replaced the "old". The urbanites and the children of the first and subsequent generations of rural migrants into the big Greek cities never developed such an organic, engaged, "rooted" sense of landscape in the first place. As in the case of the Middle East (see previous section), dissociation from the rural local cultures left behind, during colonial or occupational restructuring, also "undermined the ecological and cultural continuity between the city and the surrounding region, reorienting the city to the West" [24] (p. 116), to an externally imposed culture. In this way, the city was isolated from its rural hinterland, bearing an enormous landscape diversity, characteristic of the Greek (and Cypriot) territory. It may, of course, be argued [43] that, with the advent of the 21st century and conditions of affluence before the onset of "the crisis", Greek landscape conscience has started to re-emerge through the urbanites' nostalgia for the countryside and the "re-discovery" of landscape through domestic tourism. However, it may be too early to assess such trends, which have taken various new guises and forms, under the current grave socio-economic impacts of 'the crisis,' in the Greek-speaking part of the world.

5. Research Design and Implementation

For purposes of exploring high-level knowledge and employment of landscape-related issues, this research used the interview questionnaire approach to record and quantify landscape notions, uses and opinions of high-level landscape-related stakeholders from four countries of the broader region: Cyprus, Jordan, Lebanon and Greece. The interviews were administered to public entities, private entrepreneurs, academicians and NGO representatives (approximately 15 per country, see Appendix A for detailed list of stakeholders by country) and was followed by statistical data analysis of their responses to a fairly intensive hour-long interview questionnaire, including both open- and closed-ended questions. The interview sample is presented in the third column of Table 1. The method employed was semi-random sampling, as individuals selected for this research ought to have had extensive involvement in landscape research or planning/management. Thus, the stakeholder selection was based on prior experience or expert knowledge regarding the landscape, generally speaking; most of these stakeholders had been involved in landscape-oriented research projects and had presented considerable interest in landscape issues in their country. The content validity was established based on the rule of full landscape domain coverage, whereby all relevant aspects of landscape acknowledgement and intervention were sufficiently addressed, through an adequate number of questions. The full stakeholder survey (61 questionnaires in total), undertaken in the four countries at the outset of the ENPI-MED European-funded MedScapes Project (www.enpi-medscapes.org), included 29 questions setting the ground for the sought-out information; out of these only 4 (four) will be presented, analyzed and discussed for the purposes of this study, as follows:

- "Is the concept of Landscape officially acknowledged by your Institution/Organization?"
- "Has your Institution/Organization ever been involved in the management, planning, recording or protection of Landscape?"
- "To what degree do you consider your country's existing legal and policy framework adequate for the management of your region's/country's Landscape?"
- "What do you consider to be the pressure factors and future threats to your country's landscape?"

Table 1. Word groups used for the text analysis process of question 4.

Group ID	Keywords	*n*
1	Development	26
2	law, legal, policy, specifications, regulations, penalties, enforcement, control	19
3	urbanization, urban, construction, settlement, built-up, residential, real estate	19
4	degradation, destruction, pollution, waste, carrying capacity, limitation, erosion, quality, sensitivity	16
5	planning, plan, strategy, unplanned, off-plan, design	16
6	management, conservation, heritage	12
7	Climate	7
8	awareness, understanding, appreciation, concern	6
9	natural resources, reserves	6
10	recreation, tourism	6

The analysis of the questionnaire results required the use of a relational database system, in order to structure and recode the data (answers) for further analysis. The statistical analysis of the results was conducted with SPSS and R statistical language and was based in methodologies of descriptive statistics and illustrative graphs to present the opinions of the stakeholders. The analysis provided a series of insights presented in the final project report (http://www.enpi-medscapes.org/images/Deliverable/MEDSCAPESREPORTONSTAKEHOLDERSNEEDSKNOWLEDGE.pdf), as regards landscape-related stakeholders' assessment of the state of their national landscapes (threats, pressures, potential) and national landscape policy; cultural and natural heritage, as well as landscape acknowledgment and protection by their agencies/institutions; knowledge and employment of LCA and landscape mapping practices in landscape planning, protection and management; and appropriate stakeholder opinions, positions and intervention concerning all of the above. All of the latter will be only be drawn from in this article—as well as responses to open-ended questions pertaining to our research aims—in order to elucidate and enrich the context of our discussion. Delimitations of the undertaken survey include its relatively small sample size and its purposive nature, tailored to the requirements and constraints of the project, nonetheless rendering the outcomes of this research mostly indicative, necessitating further future research, on a broader and more representative population basis.

6. Research Findings and Analysis

Next, we present the results of this research survey, after processing the answers in SPSS and R statistical language, at both country and stakeholder levels. The stakeholder levels employed in this analysis were three: public entities, private entities and academic entities, reflecting the type of entity (Institution/Organization) represented by our interviewees.

Question 1. "Is the concept of Landscape officially acknowledged by your Institution/Organization?"

The aim of this question was to identify the official utilization of the concept of Landscape by the Institution or Organization participating in the study, through their surveyed representative. As may be seen in Figure 1, answers were generally divided more or less equally between "yes" and "no", with the exception of Greek representatives who mostly responded positively-namely that their Organizations or Institutions officially acknowledged the landscape concept. However, Cyprus respondents seemed to have the best knowledge of such official acknowledgment by their Institutions/Organizations, followed by Greek respondents. The representatives least in-the-know, as concerns this question, were those of Lebanon (with most "don't know" answers offered), rendering the Lebanese sample's responses overall rather tentative and more unreliable, as compared with those of representatives from other countries.

Figure 2 shows the absolute frequency of answers to the same question, by stakeholder type (public entity, private entity, or academic entity). In all questions regarding landscape acknowledgment and management in our study, analysis of our findings by stakeholder type indicate a very similar distribution of responses ("yes"-"no"-"don't know") among the three types of stakeholders: a rather regular pattern of tentative, yet—as expected—mostly positive responses. This fact of this recurrent accordance in the distribution of answers by stakeholder type strengthens the value of our findings, as concerns the actual situation vis-à-vis landscape stewardship (acknowledgment and management) in the four countries.

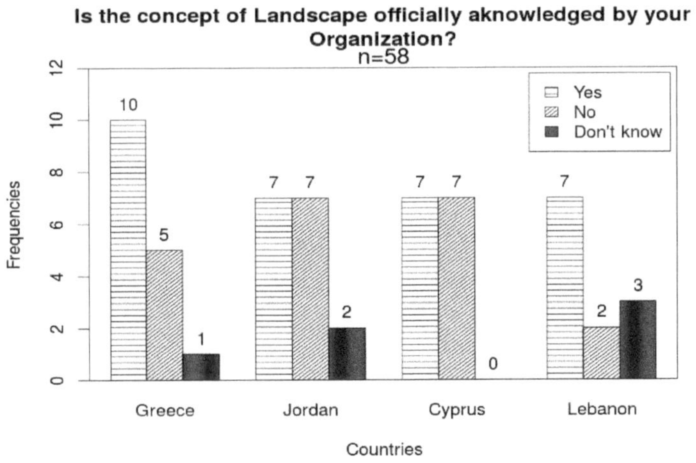

Figure 1. Frequencies of answers to the question "Is the concept of Landscape officially acknowledged by your Institution/Organization?" by country of residence.

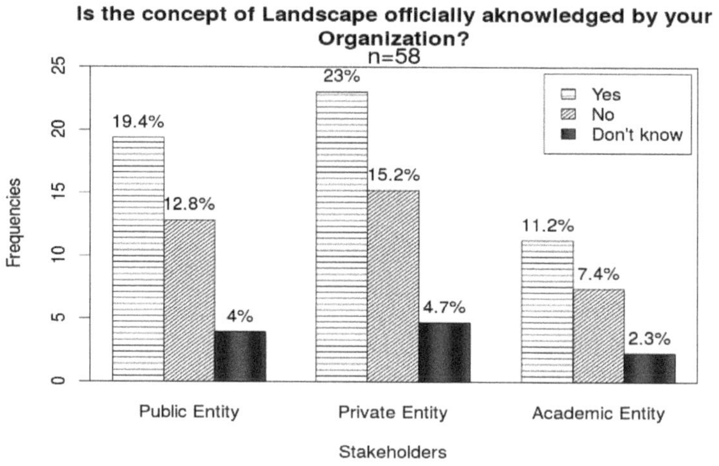

Figure 2. Frequencies of answers to the question "Is the concept of Landscape officially acknowledged by your Institution/Organization?" by stakeholder type.

Question 2. "Has your Institution/Organization ever been involved in the management, planning, recording or protection of Landscape?"

The aim of question 2 was to establish the ways of involvement of the surveyed Institutions/Organizations in any types of activities regarding the Landscape, i.e., management, planning or protection at the landscape level. The answers are of particular interest. As can be seen in the following Figure 3, all interviewees from Cyprus belong to an Institution or Organization, which has been involved in landscape-related activities, as opposed to Jordan interviewees, who indicated the lowest involvement in such tasks by their Organization/Institution (almost equal positive and negative answers). Although, obviously, the purposive selection of our interviewee sample may have affected our findings, all of our interviewees seemed to be knowledgeable (no "don't know" answers) in this regard, a positive outcome strengthening the rest of our findings. The general indication was that for all countries, besides Cyprus, there is still a ways to go, as far as landscape acknowledgment and management are concerned.

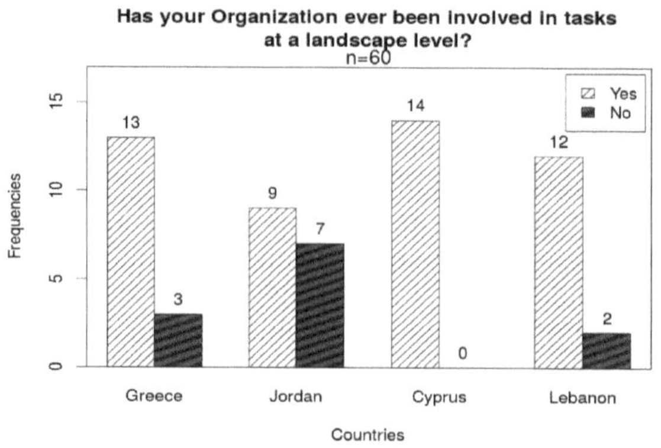

Figure 3. Frequencies of answers to the question "Has your Organization ever been involved in tasks at a landscape level (management, planning, recording or protection)?" by country of residence.

Figure 4 depicts interviewees' answers to the same question, by stakeholder type. Here, as for Figure 2 (and Figure 6, further down), our data analysis by stakeholder type indicates a very similar distribution of responses ("yes"-"no"-"don't know") among the three types of stakeholders, with a predominance again of positive answers but also illustrating accordance in their opinions, thus further strengthening the rest of our results.

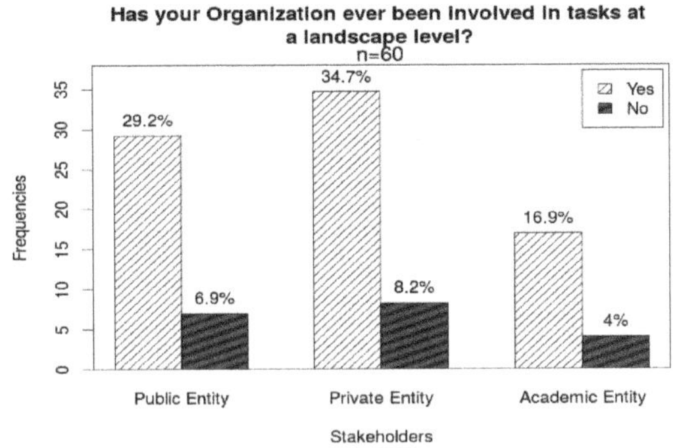

Figure 4. Frequencies of answers to the question "Has your Organization ever been involved in tasks at a landscape level (management, planning, recording or protection)?" by stakeholder type.

Question 3. "To what degree do you consider your country's existing legal and policy framework adequate for the management of your region's/country's Landscape?"

This question aimed at identifying the extent of legal/policy coverage of landscape management, by country. The results, presented in Figure 5, show that, according to the opinions of our interviewees, there is low or very low legal and policy coverage for landscape management. Thus, in all surveyed countries, the dissatisfaction with the legal/policy framework on landscape is obvious and strong, except for Cyprus, where the opinions of our interviewees seem to be divided. Cyprus was the only country were high satisfaction was quoted. In both Middle-Eastern countries, stakeholders rated their country's landscape legal/policy framework as very low or non-existent. Greece presents an intermediate case, where the highest degree of intermediate satisfaction among countries was reported but still remains one of the most dissatisfied countries with the issue under investigation.

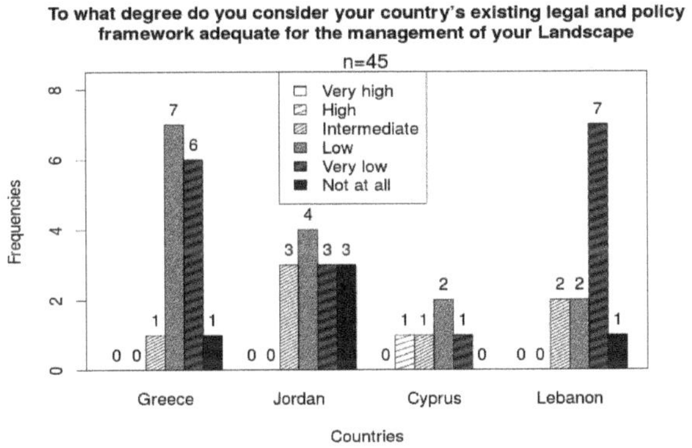

Figure 5. Frequencies of answers to the question "To what degree do you consider your country's existing legal and policy framework adequate for the management of your (region's/country's) landscape?" by country of residence.

In Figure 6, our data analysis by stakeholder type produced the same type of response distribution as in the previous two questions (Graphs 2 and 4) and therefore adheres to the same description and interpretation as that offered for these latter two graphs, above.

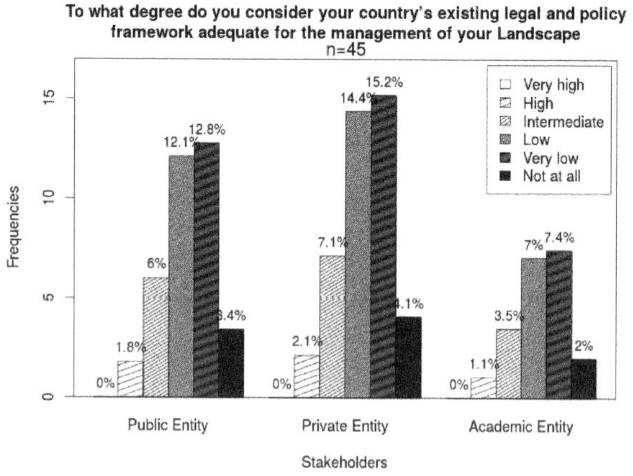

Figure 6. Frequencies of answers to the question "To what degree do you consider your country's existing legal and policy framework adequate for the management of your (region's/country's) landscape?" by stakeholder type.

Question 4. "What do you consider to be the pressure factors and future threats to your country's landscape?"

The aim of question 4 was to identify pressure drivers and future threats on the landscape, at the country level. As indicated by our findings (Figure 7), individuals from Greece and Cyprus stated that, generally speaking, "development" first (and very secondarily "recreation") were by far the most important pressure factors for the Greek and Cypriot landscapes, as opposed to the other two case studies (Lebanon and Jordan). Representatives of all case studies quoted "legal/policy/regulation" issues, "urbanization" and "degradation" as significant pressure factors impinging on their countries' landscapes. "Planning" and "management" problems were especially quoted as problems and pressure factors by the Greek respondents, more so than by any other. All of the rest of threats and pressures were quoted to a lesser, though variable, degree by all four country representatives. It is very interesting to note that "the climate" seemed to be more of an issue for the Middle-Eastern countries, as reported by their representatives, than by the European countries of our sample.

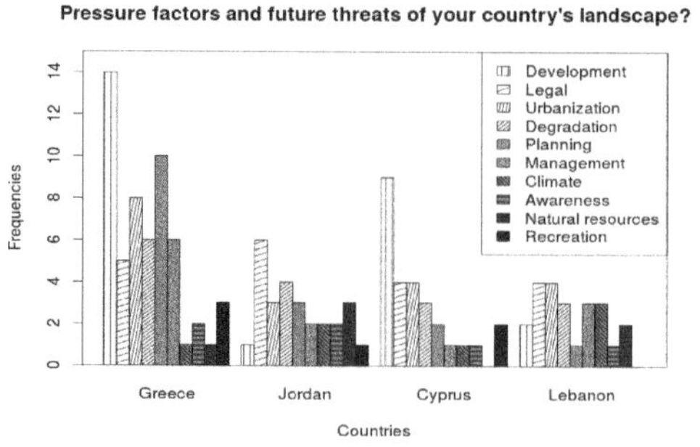

Figure 7. Frequencies of answers to the question "What do you consider to be the pressure factors and future threats of your country's landscape?" by country of residence.

Finally, Figure 8 illustrates the same data, analyzed by stakeholder type. Here, again, as for the previous questions, similar findings in terms of landscape threats and pressures were offered by all three types of stakeholders. The main exception here were the academics, who did not rate "development" as highly as private and public Institution/Organization stakeholders did, as a threat to the landscape—they rather quoted "natural resource" preservation as an area of future landscape threat or concern, together with relevant "legal and policy" issues. The latter finding may simply have to do with the academics' general orientation to and preoccupation with their own more specific scientific interests. Otherwise, there was, once again, general accordance between the answers of the three stakeholder types.

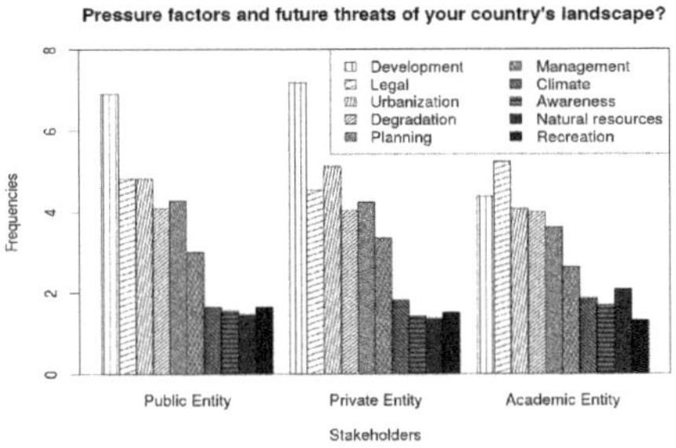

Figure 8. Frequencies of answers to the question "What do you consider to be the pressure factors and future threats of your country's landscape?" by stakeholder type.

7. Discussion of Findings

What emerges from this exploratory study of the state of landscape management (including planning and policy-making) in the four countries, through the opinions and knowledge of high-level stakeholders, is a clear but rather disappointing, image of the state of affairs. All four countries are lacking in the acknowledgement and fulfillment of their landscape needs, although they all carry and employ some concept of landscape, which lies at the basis of all relevant practices, plans and policies. It comes as no surprise that Greece and Cyprus are ahead of the Middle-Eastern countries (questions 1 and 2), at least due to the fact that Greece and Cyprus have signed and ratified the ELC but seem to be slow in implementing it. Jordan may be seen to be one step ahead of Lebanon (question 1), if only because the Jordanian organization/institution representatives interviewed seemed to carry a more concerted and informed understanding and knowledge of their countries' landscape predicament—the landscape concept being, at least nominally, acknowledged in the Jordanian legislation.

More specifically, as regards the official acknowledgement of the landscape concept by the sampled organizations/institutions and their involvement with tasks at the landscape level, on the basis of these findings, we may deduce that the landscape concept seems to have gained a stronghold in most of these countries' relevant institutions and organizations, yet not to as high a degree as desirable (because of the high number of negative responses) but generally speaking, rather somewhat more so than not (questions 1 and 2). In one only case, did a Jordanian stakeholder provide the western landscape definition, as follows, "all the visible features of an area of land, often considered in terms of their aesthetic appeal." In all other cases, in both Jordan and Lebanon, such definitions revolved around resource conservation, revealing the difference between the more instrumental local conceptions of landscape, versus formal western ones. Cyprus, generally speaking, emerges as the country most advanced in this respect. Not only did the Cypriot respondents seem to have the clearest understanding of the situation but they also reported the highest levels of involvement by their organizations/institutions in landscape-related activities (i.e., landscape recording/mapping, planning, protection, management etc.).

As regards the degree to which these respondents considered their countries' existing legal and policy frameworks adequate for the management of their country's landscapes, again, the Cypriot ones overwhelming answered positively. They, thus, put themselves at the lead concerning progress achieved vis-à-vis the implementation of the ELC and as compared to the rest of the countries, with Greece being next in line (question 3). Generally, however, the dissatisfaction with the legal/policy framework on landscape in all four countries was reported to be high, with the exception of Cyprus, where the overall picture was ambivalent, followed by Greece—where dissatisfaction with the existing legal/policy landscape framework was high. This finding may due to the widespread realization of Greece's lagging position vis-à-vis ELC implementation—as opposed to what is expected and as compared to other European country cases [44,45]. The latter finding is further upheld and strengthened by findings with regard to the final question (question 4), whereby Greek representatives, most of all other country representatives, referred to 'planning' and 'management' issues as landscape problems and pressure factors. This finding may be due to the need for proper planning and stewardship of the Greek landscape, being felt more urgently in Greece, which is especially lagging in these respects (and also, as compared to Cyprus) towards the implementation of the ELC and exacerbated as a consequence of the current socioeconomic crisis.

Coming, finally, to the problems and pressures threatening the landscape (question 4), the key problem faced in all four countries, as in the wider Eastern Mediterranean region, is the rapid pace of change currently taking place in their landscapes, jeopardizing their ecological diversity, visual quality, cultural patterns and functional integrity. Specifically, in our interviewees' words, landscape problems and threats abound at all levels of landscape planning, protection and management (from the national to the local) and in all sectors of the economy, society and space, related to the landscape (rural production, urbanization, tourism, etc.) (Table 1). There are strong pressures stemming from urbanization and tourism development in certain "hotspots" and, at the same time, rural abandonment undermines the traditional activities that helped shape the rural landscape, in the first place (Figure 9). Certain recurring problems seemed to run the gamut of all our country cases: "legal/policy/regulation" issues, "urbanization" and "degradation". This finding points not only to legal/political deficiencies in catching up with the rest of the Western world vis-à-vis their landscape stewardship but also reflects the repercussions of rampant urbanization and long-standing environmental degradation, in this part of the world: "the imprint of urbanization on the landscape of the Middle East threatens ecological stability, undermines long-term environmental sustainability and erodes the regional landscape character" [20] (p. 225). Overall, however, there was some differentiation between the European vs. the Middle Eastern countries of our study (though one may not generalize these findings to the whole Middle Eastern region, see [46,47]), in this regard, as Cyprus and Greece indicated "development" and, to a lesser degree, "recreation" as significant landscape problems and pressures, in contrast to Jordan and Lebanon. Finally, as expected, the latter two countries—already extremely water-conscious—indicated a more heightened concern about "the climate", which is probably due to serious water-shortage risks implied in scenarios of climate change and global warming.

It is interesting to put these findings in the context of the conclusions of the MedScapes Project, illustrating, through both quantitative and qualitative data, the very same trends, as reported above (http://www.enpi-medscapes.org/images/Deliverable/MEDSCAPESREPORTONSTAKEHOLDERSNEEDSKNOWLEDGE.pdf). Generally speaking, as far as the implementation of landscape mapping and character assessment methods, Cyprus and Greece, again, seemed to be one step ahead of Jordan and Lebanon. As emerges from all of the interviewees' responses, central dimensions of the concept such as the visual/aesthetic perspective of the landscape, as well as the constitution of landscape itself through people's perception (ELC), are totally absent. Specifically, in answers to our open-ended questions, the definition and utilization of the landscape concept by Middle Eastern organizations and institutions was "conservation of natural resources," "conservation of the area around the archeological sites ... in the World Heritage Sites," "management of the urban green" etc., while, only at the academic level, was there some acknowledgment of the "flexibility of the landscape definition, depending on the [course] instructor … i.e., 'Landscape Design' is a project-based course, it focuses on site-analysis … [landscape architecture rules] … as well as the development of landscape design in Far Eastern, European and Islamic cultures." Nonetheless, there was a detailed recording and addressing of the basic landscape elements (natural and cultural), pointing not only to the inclusivity of the landscape term itself in Greek and in Arabic—as acknowledged by those surveyed—but also to the rich linguistic traditions of the four countries, as regards the landscape concept, which is not easily translatable from one language to the other. Whereas, as already discussed earlier in this article, there exists one single term signifying the landscape in Greek ('topio'), this is not the case in Arabic, where, in order to connote this concept, either certain aspects of the landscape or more descriptive terms tend to be employed. The problem of translation, thus, is linked to the problem of landscape conceptualization, among the four countries and the two languages (Greek and Arabic).

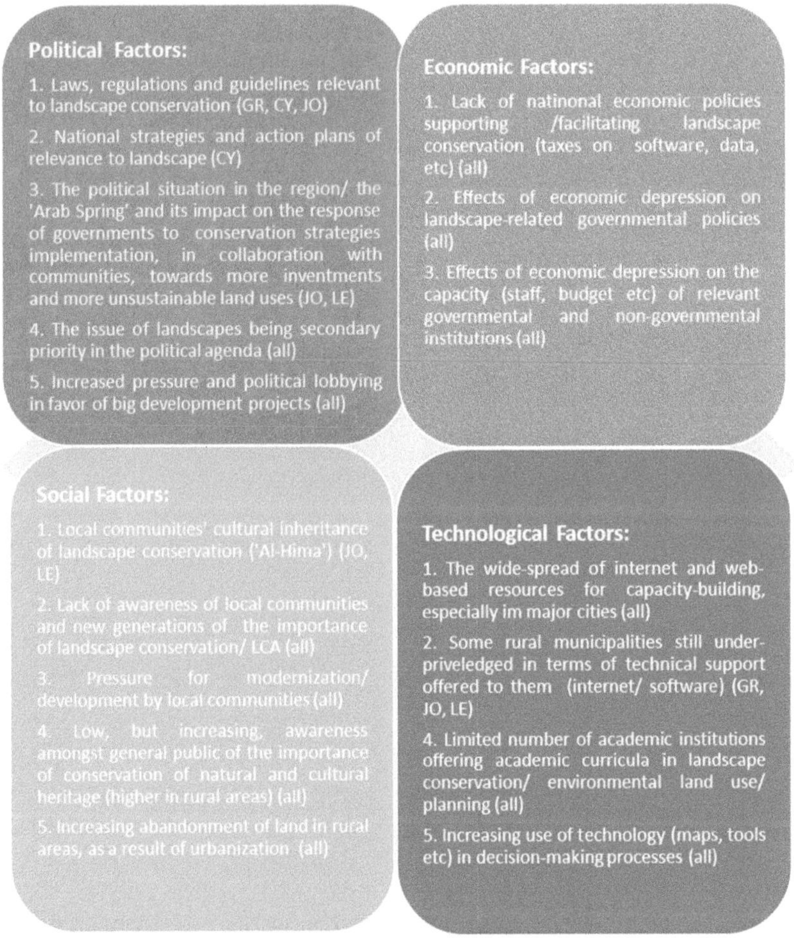

Figure 9. PEST analysis of factors impinging on landscape stewardship in the four case study countries (Cyprus, Greece, Jordan, Lebanon). Source: MedScapes final report (http://www.enpi-medscapes.org/images/Deliverable/MEDSCAPESREPORTONSTAKEHOLDERSNEEDSKNOWLEDGE.pdf).

External risk factors include political unrest and economic instability (Lebanon, Greece), difficulty of obtaining certain spatial data (Lebanon, Greece and Jordan), possible lack of state/authority willingness to consider a landscape character framework for decision making (in Lebanon and Jordan, not adhering to the European Landscape Convention). Currently, there seem to be emerging increasing trends towards holistic and sustainable conservation and management of natural and cultural resources and initiatives towards integrated land use planning. All of the latter are greatly affecting or encompassing landscape issues and causes, despite various drawbacks and threats to landscape planning, management and protection, generally speaking: (a) lack of internationally-corresponding terms and concepts as concerns the landscape; (b) lack of awareness especially among public stakeholders of the importance of landscape protection, planning and management and of a lay landscape conscience; (c) different or insufficient policies and legal frameworks regarding the landscape or landscape aspects/resources, despite common pressures on it; (d) political and planning pressures that go against the sustainable management of the landscape and natural/cultural resources, in favor of big development projects; (e) the complexity of landscape issues making its incorporation in policy-making, research agendas and higher educational programs difficult and (f) lack of accessible resources rendering the issue of landscape not a priority in the agenda of policy-makers and various other relevant stakeholders.

Conclusively, our survey findings reveal great commonalities among the Greek- and the Arabic-speaking countries of our study, as concerns the pressures currently exerted and the threats admonished vis-à-vis the landscape (Figure 1), although equally significant differences (both within and between countries) persist in the ways these pressures and problems are dealt with. As expressed in the interviewees' responses, these trends are a striking evidence of great stakeholder needs and concerns, as regards all aspects of the landscape, in all four countries—including the use of methods and techniques for its assessment, mapping, planning and management. External pressures obviously vary from NE MED sub region to sub region. On the one hand, the volatile political situation and 'the Arab spring' have brought about re-prioritizations of political and social agendas but, on the other, the effects of economic depression and threats of environmental/climate change have equally applied to the whole SE Mediterranean region: on governmental policy-making, on institutional capacity and on land use and resource planning and management. People's connection to the land and to landscape, however, is uneven, both in terms of time-place and in terms of cultural context. In the Greek world, landscape conscience is presently deficient or often non-existent, whereas in the Arab world (areas under study) such a conscience has traditionally been tied to the sustainable management of common lands (i.e., 'Al Hima'—[1]). The findings of this stakeholder survey more generally speaking, point to the need, not only for a technical approach to landscape assessment and mapping but also to the need to train and sensitize the societies in all four countries, in landscape matters, landscape interventions, as well as to raise their awareness and mobilization vis-à-vis their landscapes. Comments received through open-ended questions, in Jordan, exposed this concern as follows: "The public lack knowledge and the authorities are neglecting the future needs of the population," "the concept of landscape is vague to the general public here ... need to raise awareness of the definition and importance of the natural and cultural heritage."

---

1 'Al Hima': Hima ('protected area' in Arabic) is a community-based approach presently employed in the Middle East for the conservation of sites, species and habitats, managed by people, in order to achieve the sustainable use of natural resources. It originated more than 1500 years ago and spread along the Arab Peninsula, as a 'tribal' system for sustainable management of natural resources. It evolved further with Islam, adding ethical values and rules to the concept. It has been applied as a system for organizing, maintaining, regulating and utilizing natural pasture and rangelands, in ways compatible with ecological balance and local practices, whereby Hima management and decisions are made by the local communities themselves.

8. Concluding Remarks

The past twenty years or so have witnessed a turn more in favor of a landscape approach also in land stewardship and management. It is increasingly advocated that a whole-landscape approach provides the most appropriate spatial framework for protection and sustainable management of natural and cultural heritage; even nature conservation and management are now widely considered to be effective only when carried out at the landscape level [14,46–50].

Such a development stems not only from obligations incurred by the adoption of the ELC but also by practical necessities, prompting the development of landscape tools, methods, typologies and mapping at a variety of scales, from local to Europe-wide [44,45,51]. However, as the 'People's Convention,' according to one of its creators, Michael Dower, underscores, its emphasis is first on people, people's perceptions and on socio-cultural contexts, requiring further scientific investigation.

In this light, this study explored human ways of interrelating with landscape, at the interface between 'East' and 'West,' that is, between the Arab and the Greek worlds, in the Eastern Mediterranean region. This task was undertaken in a region in turmoil, given continuing risks from development pressures and current 'crises,' high diversity at the landscape level and the fact that the diverse and complex cultural and natural Mediterranean heritage is almost always inextricably entwined with and affected by human activities (and indeed often dependent upon them) [40], [52] (p. 299), [53]. The integrity of Mediterranean landscapes is currently threatened by common problems: pollution, overgrazing and tourism development. Lack of public awareness, political commitment or intersectoral co-operation also hinder the protection of natural heritage. Despite the common nature and origin of these problems, the actions and approaches employed so far seem to vary widely between—and even within—these worlds and countries of the region.

In conclusion, our survey findings reflect the uneasiness and confusion of a forced co-existence, in the area under study, of Western with autochthonous landscape concepts and ways of appropriation. On the one hand, then, in our interviewees' responses, we encounter the territorial/legal/commodified or visual/scenic/distantiating means of human-landscape interrelating, while, on the other hand, we discern the Arabic and ('traditional') Greek experiential ways of associating with and assessing the world, through all human senses and common everyday practices. The understanding and stewardship of any 'landscape,' such as the ones under study here, in the Eastern Mediterranean region, reflect what different societies and individuals perceive and imbue with value in their surroundings and how such perceptions and understandings of space may relate to common needs and interests, as well as collective identity formation. Therefore, such understanding and stewardship ought to take shape on their own terms, within their cultural context, irrespective of Western norms, i.e., embracing the formalistic and visual/aesthetic Western ideas of landscape, in order to uphold the great wealth of the world's concepts of landscape [20] (p. 218), [21].

While currently the scope of new meanings to the term landscape is still expanding, in some parts of the world, elsewhere part of the variety of existing terms expressing a number of different aesthetic and environmental/territorial concepts is gradually being lost, in a continuous flow of global mono-cultural colonization. However, no party's perceptions, understandings and associations with space and landscape ought to be left out in landscape protection, planning and management [1,44,45]. In sum, we ought to be making use of constructivist notions of landscape [54,55], in recovering not only the "substantive meaning" [7] of the term 'landscape' but also the great wealth of the word's landscape-related concepts, at large. For that purpose, we ought to be combining and linking such concepts as deriving from all perspectives: (a) (disciplinary and trans-disciplinary) scientific/academic; (b) (local and international) professional and (c) the civil society at large, including the public's and all relevant stakeholders' viewpoints into 'landscape' discourse, cause and practice.

Acknowledgments: The authors wish to thank Ioannis Vogiatzakis, at the Environmental Conservation and Management Programme, Open University of Cyprus, for his critical reading of segments of this paper and particularly in reference to the case of Cyprus. A part of this work has been funded from ENPI-CBCMED, for the project: MESCAPES. We would like to thank the research team of MEDSCAPES for providing data and the authorities who responded to the questionnaire.

Author Contributions: The authors wish to thank Ioannis Vogiatzakis, at the Environmental Conservation and Management Programme, Open University of Cyprus, for his critical reading of segments of this paper and particularly in reference to the case of Cyprus. A part of this work has been funded from ENPI-CBCMED, for the project: MESCAPES. We would like to thank the research team of MEDSCAPES for providing data and the authorities who responded to the questionnaire.

Conflicts of Interest: The authors declare no conflict of interest.

Appendix A. Detailed List of Stakeholders by Country

Greece

1. Ministry of Environment, Energy and Climate Change, Department of Regional Planning and Urban Development
2. Ministry of Environment, Energy and Climate Change, Special Secretariat for forests
3. Ministry of Culture and sports, Directorate of Modern and Contemporary Architectural Heritage
4. WWF Greece
5. Society for the environment and Cultural Heritage
6. Monumenta.org
7. Panhellenic Association of Landscape Architects
8. Lesvos Development Administration of the Aegean
9. Lesvos Regional Unit, Directorate of Agricultural economics
10. Region of Epirus
11. Lesvos Local Development Company
12. The natural History Museum of Lesvos
13. Nautilus Action NGO
14. Epirus Northern Pindos National Park Management Body
15. Epirus S.A. (Public Body)
16. University of the Ioannina
17. Pindos Perivallontiko NGO

Cyprus

1. Town Planning Department
2. Cyprus Architectural Heritage Organisation
3. Philippou and Associates (architects of very big projects)
4. Cyprus Sustainable Tourism
5. Forestry Department
6. Neapolis University of Paphos, School of Landscape
7. Department of Archaeology
8. Cyprus University of Technology
9. Cyprus Association for Cultural and Special Interest Tourism
10. Cyprus Tourism Organisation
11. Town Planning Department
12. Department of Environment
13. Frederick University
14. Cyprus Tourism Organisation
15. Cyprus Architects Association

Jordan

1. Ministry of Municipalities
2. Ministry of Environment
3. Jordan Development Zones JDZ
4. Aqaba Special Economic Zone Authority ASEZA
5. The Royal Scientific Society RSS
6. The Royal Society for the Conservation of Nature RSCN
7. Balqa Applied University
8. Balqa Applied University
9. United Nations Environment Program
10. German Jordanian University GJU
11. Yarmouk University
12. Jordan University for Science and Technology JUST
13. Ministry of Tourism and Antiquities
14. German Jordanian University GJU
15. Greater Amman Municipality GAM
16. Yarmouk University

Lebanon

1. Council for Development and Reconstruction (CDR), Senior Architect Urban Planner
2. Lebanese University Instructor
3. Tourguide syndicate Lebanon, Program manager In NGO Sector and trainer
4. Association for Forests, Development and Conservation, Project Coordinator
5. University of Balamand, Director Archaeology and Museology
6. Ministry of Education, Coordinator in Environmental Education
7. Prime Ministry Cabinet, General Director of Technical Affairs
8. The Other Dada, Founder and Lead Architect at theOtherDada
9. UNDP/MOE, Project Manager
10. Lebanon Green Building Council, Vice President and Board Member
11. Al-Shouf Cedar Society, Environmental Awareness Coordinator
12. The Lebanese Environment Forum, Civil Engineer
13. Al AKhbar Newspaper, Reporter
14. ASAMCO–fertilizers and pesticides company, Manager, Agricultural Engineer
15. Khatib & Alami Co.

References

1. European Landscape Convention (ELC). Committee of the Ministers of the Council of Europe, European Landscape Convention and Reference Documents. In *European Landscape Convention and Reference Documents, Council of Europe*; Council of Europe: Strasburg, Germany, 2008; Volume 6.
2. Drexler, D. *Landschaft und Landschaftswahrnehmung: Untersuchung des Kulturhistorischen Bedutungswandels von Landschaft Anhand eines Vergleichs von England, Frankreich, Deutschland und Ungarn*; Technische Universität München, Lehrstuhl für Landschaftsökologie: München, Germany, 2010.
3. Ueda, H. *A Study on Resident Landscape Perception through Landscape Image: Four Case Studies in German and Japanese Rural Communities*; Architektur, Stadtplanung, Landschaftsplanung der Universität Kassel: Kassel, Germany, 2009.
4. Bruns, D.; Zhang, K.; Zhao, J. Begriff der Landschaft in China. In *Landschaften: Theorie, Praxis und Internationale Bezüge. Schwerin*; Oceano Verlag Schwerin e.K.: Schwerin, Germany, 2013; pp. 133–152.
5. Bruns, D.; van den Brink, A. 'Pays'-'Land'-'Yuan Lin'. The power of landscape (architecture) terms. In Proceedings of the ECLAS 2012 Conference, Warsaw, Poland, 19–22 September 2012; pp. 447–449.

6. Cosgrove, D.E. *Social Formation and Symbolic Landscape*; The University of Wisconsin Press: Madison, WI, USA, 1998.
7. Olwig, K.R. Recovering the substantive nature of landscape. *Ann. Assoc. Am. Geogr.* **1996**, *86*, 630–653. [CrossRef]
8. Jackson, J.B. *Discovering the Vernacular Landscape*, 3rd ed.; Yale University Press: New Haven, CT, USA, 1984; Volume 3.
9. Meinig, D.W. The beholding eye: Ten versions of the same scene. In *The Interpretation of Ordinary Landscapes: Geographical Essays*; Oxford University Press: Oxford, UK, 1979; pp. 33–48.
10. Rose, G. *Geography as a Science of Observation: The Landscape, the Gaze and Masculinity*; Historical Geography Research Group of the Institute of British Geographers: London, UK, 1992; Volume 28.
11. Nash, C. Reclaiming vision: Looking at landscape and the body. *Gend. Place Cult. J. Fem. Geogr.* **1996**, *3*, 149–170. [CrossRef]
12. Appleton, J. *The Experience of Landscape*; Wiley Chichester: Chichester, UK, 1996.
13. Cosgrove, D. Cultural landscapes. In *A European Geograph*; Unwin, T., Ed.; Addison Wesley Longman Limited: New York, NY, USA, 1998; pp. 65–81.
14. Brunetta, G.; Voghera, A. Evaluating landscape for shared values: Tools, principles, and methods. *Landsc. Res.* **2008**, *33*, 71–87. [CrossRef]
15. Cosgrove, D.; Daniels, S. *The Iconography of Landscape: Essays on the Symbolic Representation, Design and Use of Past Environments*; Cambridge University Press: Cambridge, UK, 1988; Volume 9.
16. Lukermann, F. Geography as a formal intellectual discipline and the way in which it contributes to human knowledge. *Can. Geogr. Géogr. Can.* **1964**, *8*, 167–172. [CrossRef]
17. Terkenli, T. Understanding and Analyzing Cultural Landscapes. In *Perspectives on Cultural Landscapes of the Maltese Islands*; Conrad, E., Cassar, L.F., Eds.; International Environment Institute, University of Malta: Msida, Malta, 2008.
18. Lorimer, H. Cultural geography: The busyness of beingmore-than-representational. *Prog. Hum. Geogr.* **2005**, *29*, 83–94. [CrossRef]
19. Ingold, T. The temporality of the landscape. *World Archaeol.* **1993**, *25*, 152–174. [CrossRef]
20. Makhzoumi, J.M. Landscape in the Middle East: An inquiry. *Landsc. Res.* **2002**, *27*, 213–228. [CrossRef]
21. Noparatnaraporn, C. Living place and landscape in Bangkok: The merging character. In Proceedings of the Hawaii International Conference on Arts and Humanities, Kapolei, HI, USA, 16–18 October 2003.
22. *Landscape: Politics and Perspectives*, 1st ed.; Bender, B. (Ed.) Bloomsbury Academic: London, UK, 1993.
23. Duncan, J.; Duncan, N. (Re) reading the landscape. *Environ. Plan. D Soc. Space* **1988**, *6*, 117–126. [CrossRef]
24. Hokema, D.; Koura, H.; Jiraprasertkun, C.; Makhzoumi, J. International Concepts of Landscapes, Theory Basis. In *Landscape Culture-Culturing Landscapes*; Springer: Berlin, Germany, 2015; pp. 67–126.
25. Cosgrove, D. Symbolic landscape and social formation. *Lond. Croom Helm* **1984**. [CrossRef]
26. Tuan, Y.-F. Thought and landscape: The eye and the mind's eye. In *The Interpretation of Ordinary Landscapes*; Meinig, D.W., Ed.; Oxford University Press: Oxford, UK, 1979; pp. 89–102.
27. Barakat, H. *The Arab World: Society, Culture, and State*; University of California Press: Berkeley, CA, USA, 1993.
28. Lehrman, J.B. *Earthly Paradise: Garden and Courtyard in Islam*; University of California Press: Berkeley, CA, USA, 1980.
29. Pungetti, G.; Makhzoumi, J. Island landscape and seascape yet to come. In *Island Landscapes: An Expression of European Culture*; Routledge: Abingdon, UK, 2016; p. 289.
30. Terkenli, T.S. Landscape research in Greece: An overview. *Belgeo. Rev. Belge Géogr.* **2004**, *2*, 277–288. [CrossRef]
31. Vogiatzakis, I.; Trigkas, V.; Zomeni, M. Landscape identity, character and governance in Cyprus. In *Island Landscapes: An Expression of European Culture*; Pungetti, G., Ed.; Routledge: New York, NY, USA, 2016; pp. 168–175.
32. McNeill, W.H. *The Metamorphosis of Greece since World War II*; Blackwell: Oxford, UK, 1978.
33. Pettifer, J. *The Greeks: The Land and People since the War*; Penguin Books: London, UK, 2000.
34. Vogiatzakis, I.; Manolaki, P.; Trigkas, V. *LCA Training and Implementation*; Medscapes WP5 Final Report; Penguin Books: London, UK, 2015.
35. Terkenli, T.S. In Search of the Greek Landscape: A Cultural Geography. In *The European Landscape Convention*; Jones, M., Stenseke, M., Eds.; Springer: Dordrecht, The Netherlands, 2011; pp. 121–141.

36. Terkenli, T.S.; Tsalikidis, I.; Tsigakou, F.M. The physical landscape of Greece in nineteenth-century painting: An exploration of cultural images. In *Art and Landscape*; Anagnostopoulos, G., Ed.; Panayotis and Effie Michelis Foundation: Athens, Greece, 2001; pp. 618–632.
37. Stathatos, J. *The Invention of Landscape: Greek Landscape and Greek Photography, 1870–1995*; Camera Obscura: Thessaloniki, Greece, 1996.
38. Lambropoulos, V. *The Rise of Eurocentrism: Anatomy of Interpretation*; Princeton University Press Princeton: Princeton, NJ, USA, 1993.
39. Kambouridis, H. *Modern Greek Landscape Painting from the 18th to the 21st Century: Vision, Experience and Reconstitution of Time*; Foundation of Visual Arts and Music V. & M. Theocharakis: Athens, Greece, 2009.
40. Rackham, O.; Moody, J. *The Making of the Cretan Landscape*; Manchester University Press: Manchester, UK, 1996.
41. Manolidis, K. I Ypaithros kai ta Traumata tis Metapolemikis Elladas. In Proceedings of the Scientific Meeting I Diekdisisi tis Ypaithrou, Volos, Greece, 28–30 March 2008.
42. Olwig, K.R. Landscape as a contested topos of place, community, and self. In *Textures of Place: Exploring Humanist Geographies*; Adams, P.C., Hoelscher, S., Till, K.E., Eds.; University of Minnesota Press: Minneapolis, MN, USA, 2001; pp. 93–117.
43. Terkenli, T. Development of Landscape Character Assessment as a tool for effective conservation of natural heritage in the Eastern Mediterranean—Draft stakeholders needs-assessment questionnaire analysis and first results (WP4). In Proceedings of the Meeting of Experts, Limassol, Cyprus, 28–30 March 2014.
44. De Montis, A. Impacts of the European Landscape Convention on national planning systems: A comparative investigation of six case studies. *Landsc. Urban Plan.* **2014**, *124*, 53–65. [CrossRef]
45. Conrad, E. The European Landscape Convention: Challenges of Participation. *Landsc. Res.* **2012**, *37*, 383–385. [CrossRef]
46. Perevolotsky, A.; Sheffer, E. Integrated management of heterogeneous landscape—Mediterranean Israel as a study case. *Isr. J. Ecol. Evol.* **2011**, *57*, 111–128. [CrossRef]
47. Marignani, M.; Chiarucci, A.; Sadori, L.; Mercuri, A.M. Natural and human impact in Mediterranean landscapes: An intriguing puzzle or only a question of time? *Plant Biosyst.* **2017**, *151*, 900–905. [CrossRef]
48. Griffiths, G.H.; Vogiatzakis, I.N.; Porter, J.R.; Burrows, C. A landscape scale spatial model for semi-natural broadleaf woodland expansion in Wales, UK. *J. Nat. Conserv.* **2011**, *19*, 43–53. [CrossRef]
49. Woodcock, B.A.; Vogiatzakis, I.; Westbury, D.; Lawson, S.C.; Edwards, R.A.; Brook, J.A.; Harris, J.S.; Lock, A.K.; Maczey, N.; Masters, J.G. The role of management and landscape context in the restoration of grassland phytophagous beetles. *J. Appl. Ecol.* **2010**, *47*, 366–376. [CrossRef]
50. Wu, J. *Landscape of Culture and Culture of Landscape: Does Landscape Ecology Need Culture?* Springer: Berlin, Germany, 2010.
51. Birks, H.J.B. Ecological palaeoecology and conservation biology: Controversies, challenges, and compromises. *Int. J. Biodivers. Sci. Ecosyst. Ser. Manag.* **2012**, *8*, 292–304. [CrossRef]
52. King, G. *A Solution to the Ecological Inference Problem: Reconstructing Individual Behavior from Aggregate Data*; Princeton University Press: Princeton, NJ, USA, 1997.
53. Braudel, F. *The Mediterranean and the Mediterranean World in the Age of Philip II*; University of California Press: Berkeley, CA, USA, 1995; Volume 2.
54. Vivien, B. *An Introduction to Social Constructionism*; Routlegde: Abingdon, UK, 1995.
55. Ermischer, G. Mental landscape: Landscape as idea and concept. *Landsc. Res.* **2004**, *29*, 371–383. [CrossRef]

land

Article

Landscape and Urban Governance: Participatory Planning of the Public Realm in Saida, Lebanon

Jala Makhzoumi * and Salwa Al-Sabbagh

UNIT44 Architecture, Landscape, Ecological Planning, Urban Design, Yassine Building, Phoenicia Street, Ain Mreiseh, Beirut, Lebanon; ssabbagh@unit44.net

* Correspondence: jalamakhzoumi@googlemail.com; Tel.: +961-3-436-300

Received: 13 March 2018; Accepted: 5 April 2018; Published: 12 April 2018

Abstract: The political shift in Lebanon since the 1990s towards market-led development has encouraged the incremental appropriation of public spaces and state lands, and their conversion into gated, monitored enclaves that serve a privileged few. The process disregards the role of the urban public realm and undermines its potential as an inclusive space and enabling platform for urban governance. This article advocates a participatory approach to urban development, one that engages local stakeholders, institutions, and the public at large as active partners working towards sustainable urban futures. We draw on a case study in Saida, Lebanon, to illustrate participatory planning methods and demonstrate the role of landscape architects in enabling community-led development that is place responsive and sensitive to local narratives of heritage and identity. The project's participatory methodology and landscape architecture's expansive framing, the paper argues, democratizes the planning process and contributes to urban governance that empowers local authorities and local stakeholders in the face of privatization and market-led development.

Keywords: participatory; governance; landscape; public realm; urban environment; local authority; Lebanon; Mediterranean

---

1. Introduction

1.1. Why Participatory Planning?

The gradual change in postwar Western democracies from the 'traditional', top-heavy conception of government, the 'direct provider' of welfare and services, towards a bottom-up, decentralized model reflects widespread belief that the ideal place for democratic training and debate is at the local level [1–3]. The latter is argued on the premise that the system of values and norms are more homogeneous at the urban, i.e., city, level than at the national one. By the early 1990s, elected local governments became an important facet of what came to be known as 'local governance'. The political shift from "government to governance"[1], favored process rather than institutions [4] (p. 5), making governance[2] a priority in the discourse of democracy in cities.

A 'participatory' approach is directly associated with the political shift towards decentralization and the changing discourse on governance. Participatory methods, whether applied to governance, development, or democracy, aim to provide opportunities for the participation of individuals and

---

1 While 'government' refers to "the formal institutional structure and location of authoritative decision-making in the modern state", 'governance' signifies new processes of governing and new methods by which society is governed, and a focus on the interdependence of governmental and non-governmental organizations working together [5] (p. 34).

2 UN-Habitat defines 'governance' as the many ways that institutions (i.e., local authorities) and individuals "organize the day-to-day management of a city, and the processes used for effectively realizing the short-term and long-term agenda of a city's development" [6].

institutions in processes that generate ideas and action for inclusive development and social change. Capacity-building is a key to 'participatory governance' and accountability is its central lever [7,8]. Accountability develops from three key lines of action [9]: to provide enabling spaces; to initiate bonding and bridging communication; and to strategize power-shifting, which, in many ways, is the outcome of the former two. Together, these strategies provide ways to engage communities and marginalized factions of society and empower them to influence local and central decision-making.

1.2. The Political Landscape of Lebanon

Lebanon is a small country of 4.5 million inhabitants in the Eastern Mediterranean. Mountains and sea account for an exceptionally picturesque and heterogeneous physical setting. Landscape heterogeneity is matched by the religious and political diversity of Lebanese society. A parliamentary democratic system set in place following independence from the French Mandate in 1943 ensures that government offices are distributed proportionately to represent the religious diversity. The Lebanese civil war (1975–1990) interrupted elections, undermined the prevailing quasi-democratic system, encouraged informal economies and laissez-faire development and the accelerated growth of towns and cities along Lebanon's 225 km long Mediterranean coast, where the largest cities are located, including the capital city, Beirut.

At the same time, highly-centralized government institutions, another colonial legacy that continued to dominate postcolonial decades, results in weak local authorities in terms of decision-making and in responding to local needs. Apart from the National Physical Master Plan for the Lebanese Territories [10], city master plans that date to the 1960s and broad planning legislation, urban growth, indeed, territorial development, remain unregulated. Urban/regional infrastructure network is undertaken by the Council for Development and Reconstruction (CDR) and the Ministry of Public Works with a minimum cooperation from local authorities, who lack financial and technical resources.

Since the 1990s, state politics have shifted gradually away from democratic rule towards neoliberal politics leading to serious repercussions. On the one hand, market-led development has meant excessive privatization by global corporations and realty developers that is changing the spatial morphology and undermining the urban ecology. Market-led development, it is argued, animates the national economy, a priority considering the political instability in Lebanon and the region. On the other hand, the role of the state, as the body upholding public rights and protecting the public realm, has been marginalized. As a result, the public realm is increasingly suffering, as state land is appropriated and transformed into exclusive development that privileges a handful at the expense of the urban majority [11]. Ecologically and spatially, too, the city has changed as open, soft landscapes are converted into artificial ones, river courses channelized and waterfronts are privatized and built-up. Spatially excluded, socially marginalized, their memories and traditional practices sidestepped, the political majority lose their voice in as a supposedly democratic government system fails. As a tool for collective decision-making and socially just development, participatory planning is one alternative to address failures of governance in Lebanese cities. Accepting participatory planning as the starting point, this article aims to explore this approach and whether landscape architects can play a role in participatory planning that can contribute to urban governance.

1.3. A Landscape Framing of a Participatory Approach

The evolving relationship between people and the environment they inhabit is at the heart of the idea of landscape and, equally, the complex relationship that embraces the tangible, physical setting, and intangibles, such as associated cultural values and social practices. A landscape framing spatializes and contextualizes abstract discourses and anchors them in a specific place and culture. It follows that landscape is increasingly applied to frame the discourse on human rights [12], address custom, law, and the body politic [13], and tackle governance and democracy [14]. Ecology and politics are similarly influences that impact the fast-evolving discourse on the landscape, bringing together modes of being, acting, and thinking that reflect the past to project into the future [15].

The landscape's evolving epistemology, in turn, expands design practice and outlook spatially while sensitizing architects, landscape architects and urban designers to local ecologies, cultural practices, and local community needs and aspirations. Humanized and culture-invested landscapes then serve as a dynamic and enabling medium for socio-economic betterment and political empowerment [11]. In theory, therefore, a landscape design framing of the public realm can encourage "inclusive and democratic, long-term and integrated, multi-scalar and territorial" development, which are prerequisites for good governance [6]. In the following sections we outline the project methodology, the logistics of a landscape framing and, thereafter, discuss the applications and findings. We conclude with a discussion of participatory practices emerging from the project reviewing their impact on urban governance in Saida.

2. Materials and Methods: Saida Case Study

2.1. Case Study Method in Landscape Architecture

Case studies in research are gaining ground and credibility as a research method that can be applied across a range of disciplines (medicine, engineering, planning) as a method that utilizes real-life examples as a source for expanding knowledge and generating scientific discourse [16]. In a report commissioned by the Landscape Architecture Foundation, Francis concludes that "a case study method is a highly appropriate and valuable approach" [17] (p. 5). Considering that landscape architecture is a new profession, i.e., in comparison to architecture and city planning, and that the body of research is somewhat limited, the case study method can "answer big questions at the intersection of policy and design" and prove "useful in participatory planning, for culturally sensitive design and for studies trying to refine or test emerging concepts and ideas" [17] (p. 13). This is especially the case in the Middle East, where the meaning of the word landscape is limited to natural scenery and the professional scope similarly narrowed to beautification [18]. It follows that the use of case studies, whether in graduate research or in professional discourse, expands the prevailing understanding of the word and contributes to better understanding of the professional scope.

In this article, we draw on the case study of a project in Saida (Phoenician Sidon), Lebanon, to elaborate the project participatory planning methodology. The article focuses on the "Landscape, Environment, and Ecology" component of the project and explores a landscape framing of the public realm, i.e., state-owned lands, such as river corridors and the seafront, to test the efficacy of landscape architecture in contributing to urban governance.

2.2. Saida City Profile

Forty kilometers south of the capital Beirut, the city of Saida[3] is the third largest city in Lebanon (Figure 1). Saida's natural rocky harbor is one of oldest harbors in the Levant, having served as the main port city for Damascus and the region up to the early twentieth century. Municipal Saida occupies a coastal plain, 1–1.5 km wide and 7 km long, delimited from the east by gentle foothills and punctuated with a series of watercourses. The Awali and Sayniq Rivers, respectively, define the northern and southern boundaries of the municipality. In between are smaller streams, seasonal watercourses, that form natural, ecological corridors and historic, cultural caravan routes connecting the desert to the Mediterranean Sea. Rivers and streams are alive in the collective memory of the city's inhabitants, a source of life, irrigating the city orchards, structuring residential neighborhoods, and shaping the urban morphology. Saida's urban footprint covers 62% of the municipal area, 7.50 km^2; the remaining

[3] With a total population of 149,600, Saida is the third largest city in Lebanon; 80,000 of Saida inhabitants are Lebanese; the rest, Palestinians.

28% is costal agriculture. Of the total estimated population, 149,600 inhabitants, 80,000 are Lebanese, while the rest are Palestinians, and Saida houses the largest Palestinian refugee camp in the Middle East. The religious diversity is typical to Lebanon, which includes a majority of Sunni Muslims, 79%, Shiite Muslims, 10%, and 7% Catholic and Maronite Christians.

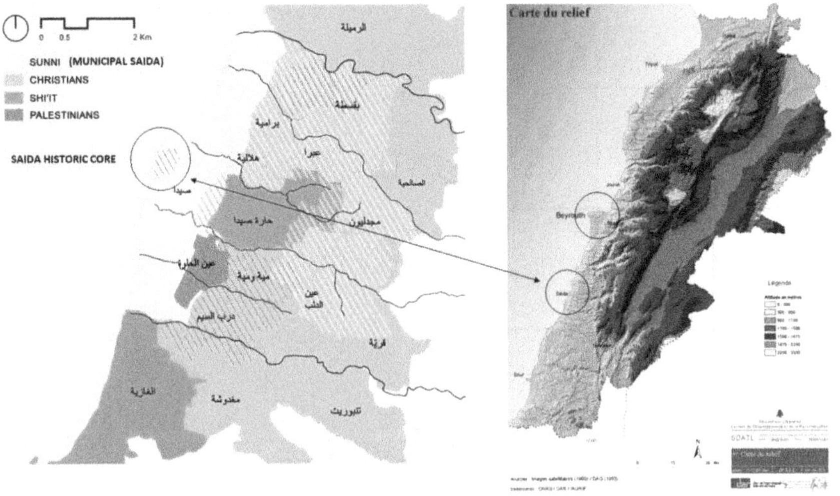

(a) (b)

Figure 1. (**a**) Saida, religious composition (source: authors) and (**b**) the location south of the capital of Lebanon, Beirut [10].

Saida was designated as a 'gate city' to Southern Lebanon [10], implying that it serves as an administrative and service portal, a position it held prior to the Lebanese Civil War. Historically, Saida's economy was maritime (commerce, fishing) and agricultural (citrus cultivation). Today both are under threat. Declining agriculture profitability has encouraged realty speculation targeting agricultural lands. Additionally, Saida's harbor is the smallest commercial port in Lebanon today and no longer viable economically. Cultural tourism is proving a new economic resource, considering the relatively intact medieval city with its abundance of ancient archaeological, Byzantine, and Ottoman sites and buildings. The significance of Saida, however, lies as well in the services it provides and its proximity to the capital, which accounts for the rapid urban growth and expansion at the expense of quality living the city was known for.

The Saida municipality, an elected body, plays a limited role in the highly-centralized development planning in Lebanon. Administrative, financial, and technical limitations restrict the municipal role in Lebanese cities, reducing local governance to nominal approval and support of projects assigned and funded by the central government. Moreover, project selection follows a political process of consensus decision-making, "a 'political game' that often reaches a complex equilibrium that distributes public spending across Lebanon" (Figure 2) [19].

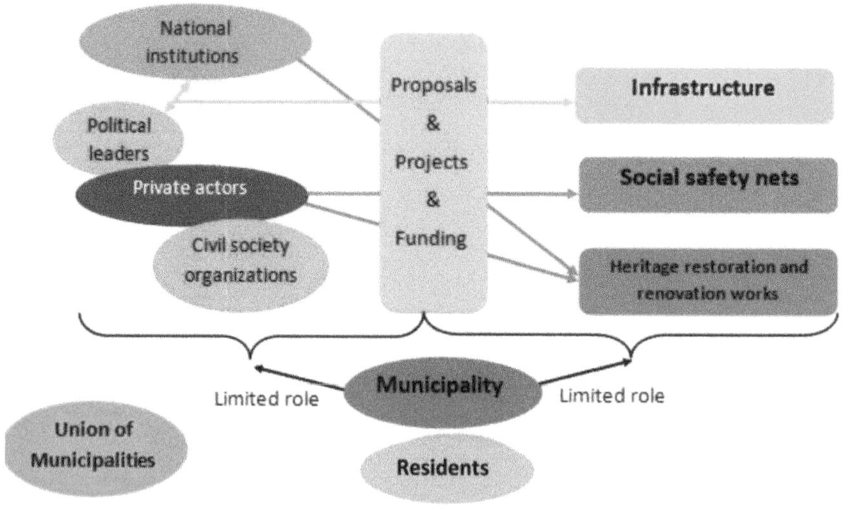

Figure 2. Schematic representing the current process of decision-making in the Saida municipality. Project proposals, their funding and implementation is highly centralized, controlled by national institutions and leadership [19].

Residents are not effectively involved in the decision-making process seeing that the municipality, itself, which is the only elected body in the city, plays a limited role in this process. Nor are decisions made with a regional development vision. The latter undermines collaboration of the municipalities that form the Saida and Zahrani Union of Municipalities (UoM) and does not allow for the UoM to address service and infrastructure projects that cross municipal boundaries and necessitate cooperation of municipalities in the union, for example, solid waste management and sewage. Sewage is dumped into rivers, watercourses and state land, causing a major environmental problem in the Saida municipality that receives the discharge downstream. The absence of a national institutional and legal framework that supports, enables, and empowers municipalities, and the prevalence of generic engineering solutions that continue to be planned with disregard to local context, have dire repercussions to environmental health and human health alike [19].

2.3. *MedCities City Development Strategy*

MedCities is an initiative created in Barcelona, in November 1991, aspiring to "strengthen decentralized actions" through technical assistance to promote awareness of urban environmental problems and use them as a platform for action that empowers municipalities in developing countries to manage urban environmental issues [20].

Members from the Saida Municipal Council applied and were considered, and eventually accepted, by MedCities for funding. The outcome is the Saida Urban Sustainable Development Strategy (USUDS) project that was launched in the fall of 2012 with funding secured from the European Council. MedCities' participatory, bottom-up methodological approach was a key to "reinforce the role, competences and resources (institutional, financial and technical) of local administrations in the adoption and implementation of sustainable local development policies" and equally, to "develop citizens' awareness and involvement in the sustainable development of their towns and cities" [20]. MedCities methodology views strategic planning as a continuous management process rather than a fixed plan, i.e., a master plan. It follows that social and economic agents, neighborhood communities, unions, entrepreneurs, and local authorities become active players in the urban vision and future development. Active involvement of all stakeholders in decision-making and the role of local authorities taking charge of financing and execution is seen by MedCities as the surest path to strengthen urban governance. In the Saida USUDS, the Saida municipality was the 'client' signing the MedCities contract, entrusted with follow-up and delivery of project final output.

3. Application of a Participatory Approach

The MedCities participatory approach was applied through all phases of the Saida USUSD[4]. Additionally, it is the broad scope of the MedCities methodology that is reflected in the interdisciplinary project tracks to which lead and supporting national expert are assigned to form the project team. The participatory approach in Saida was realized by identifying key actors whose participation was central to the analysis and formulation of a shared city vision and city development strategy. Key actors were stratified into three overlapping levels. The first embraced the local authority, realized through the "Municipal Steering Committee" composed of twelve members that represent the administrative, economic, and social sectors[5]. The second level for participation was achieved by forming working groups for each of the five disciplinary tracks[6]. The third level of participation, targeting the general public, was achieved through three town hall meetings, one at the launching of the project, the second halfway through the project[7], and a third at the end of the project. Participants on all three levels worked with the project consulting team collectively and independently.

Working groups were key to brainstorming the collective vision for the city and also representative members of the municipal council that were present in most meetings. Participation of the municipality staff was intensified in the last three phases of the project for budgeting and technical feedback. At the same time, political endorsement of, and support for, the Saida USUDS project was given by the Hariri Foundation and Deputy Bahia Hariri, both based in Saida. Endorsement, a necessary dimension for the success of development projects in Lebanon, is also problematic because of the danger that it can overshadow and dominate, rather than enforce, local governance. The multi-disciplinary team of national consultants[8], appointed by MedCities in consultation with the Saida municipality, was another cornerstone ensuring the success of the Saida USUDS project. The following excerpt outlines the team position and guiding vision that aims to expand the institutional base and decision-making process in the city (Figure 3):

> "The team adopted an institutional perspective in which participatory democracy plays an essential role. Participatory democracy is a process of deliberation based on effective cooperation between the various institutions in the city, via the creation of institutional networks that exchange information, knowledge, and experience between the municipality and the socio-economic actors in the city." [19]

---

4 The six project phases include: (1) Preparation and Launching; (2) Diagnosis and Participatory Analysis; (3) Formulation of a Shared Vision; (4) Strategy Formulation to Translate Vision into Concrete Actions; (5) Preparation of Action Plans and Estimated Budget; and (6) Implementation, Monitoring and CDS Evaluation [21].

5 The members include: the deputy mayor, four municipal council members, vice president for the Saida Traders Association, Said Welfare Society, Director of the Sea Mosque Organization, executive director from the Hariri Foundation, General Manager of the Saida Chamber of Commerce, and representative from the Lebanese University.

6 Working Group members include: 13 participants for the Urban Infrastructure track; eight for the Landscape, Environment and Ecology track; 16 for Cultural and Natural Heritage; 12 for the Employment in Traditional Industries and Trades; 14 participants for the Institutional and Legal Framework track; and 10 for the Local Economic Development.

7 The second Town Hall Meeting was cancelled because of armed confrontation between political parties on the day, which escalated to engulf the city for the following weeks.

8 Project disciplinary tracks and their respective national expert(s): "Local Economic Development", "Social Structure and Urban Governance", Jad Chaaban, Ilina Srour and Kanj Hamade; "Cultural Heritage" and team leader, Howayda Al Hanthy; "Urban Infrastructure", Omar Abdulaziz Hallaj and Giulia Guadagnoli; "Landscape Environment and Ecology", Jala Makhzoumi and Salwa Al-Sabbagh.

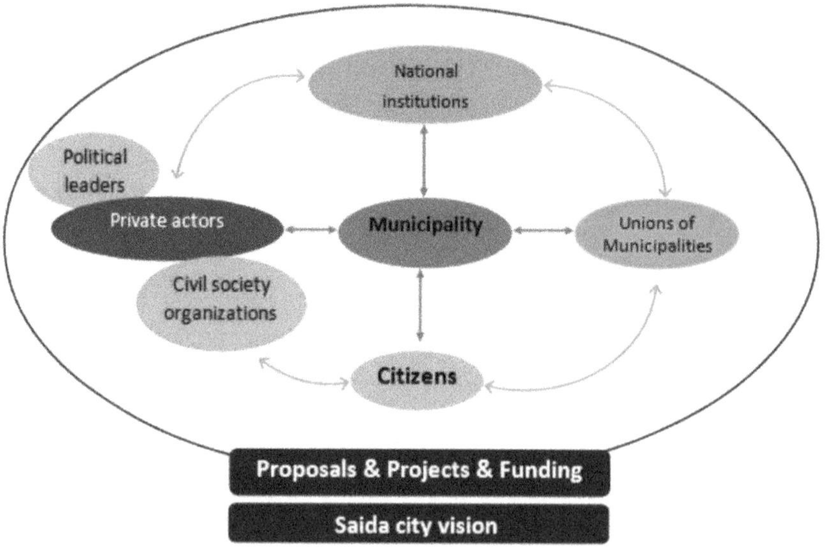

Figure 3. The decision-making process cycle that needs to be instituted to strengthen urban governance [19].

The Saida USUDS project vision, articulated at the end of Phase 3, prioritizes good governance as a foundation for quality living. The vision is supported by four overlapping, mutually-inclusive pillars: adequate health and social protection; a diversified economy; healthy environment; and protected urban heritage, both natural and cultural. The position of the project team confirms that urban governance in Saida can only proceed from accepting religious and cultural diversity and a democratic participatory approach that includes all the city's residents, Lebanese and Palestinian. The vision and strategies proposed can and should enable the city to lobby for change in the laws and regulations to allow a semblance of independent decision-making. It is "only through an institutional approach for governance can such a decision of the city be implemented" [19].

The "Landscape, Environment, and Ecology" component of the Saida USUDS was developed by applying a holistic ecological landscape design methodology [22] to take stock of existing conditions, identify environmental problems, locate natural resources, and assess their potential and limitations for sustainable urban development. The methodology enabled a landscape framing of the environment and natural resources that embraced rivers and seasonal watercourses, the Mediterranean seafront, and the abandoned railway corridor, reaffirming that they are state-owned lands, part of the 'public realm', that should be preserved from privatization and informal encroachment. The environmental health and ecological integrity of these features was a key to the landscape analysis. A holistic landscape framing as such spatializes the environmental challenge, by considering the limitations and potentials of these morphological, water-related features rather than focusing solutions to them as environmental 'problems'. A holistic landscape framing is also multifunctional in accepting that these features are, at once, an environmental resource, an ecosystem and urban landscape heritage that is culturally meaningful and integral to the collective memory of the people of Saida. The landscape vision integrates rivers, the seafront and other components identified as forming a 'Green-Blue Network' that protects them as landscape heritage, environmental resource, ecological corridors and amenity spaces[9] (Figure 4). A landscape framing of the 'public realm', as such, prioritizes environmental health and urban connectivity. Additionally, because the Green-Blue Network is spatially expansive, covering the entire municipal area, it is accessible to all the inhabitants of Saida and the neighboring municipalities, a public realm that is socially inclusive rather than being exclusive to privileged factions of the society [22]. Just as significantly, the proposed network draws on historically living cultural practices that use the rivers, coast, and railway tracks for social gathering, thus reaffirming the urban distinctiveness of Saida and countering the homogenizing approach of market-led development[10].

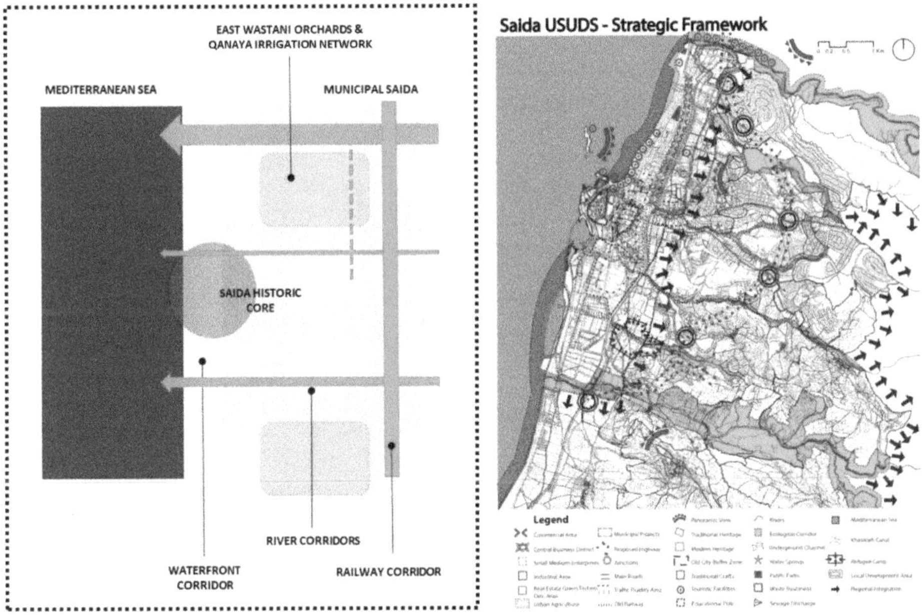

Figure 4. Saida USUDS landscape concept and consolidated structure plan embracing all five project disciplinary tracks [23].

4. Discussion

Revisiting the premise posed at the outset of this article: does an expansive landscape design framing contribute to urban governance? If yes, why and how is its contribution different from, and complementary to, the participatory approach of the Saida USUDS project? Drawing on the participatory and transdisciplinary approach of the MedCities as applied in Saida, we posit that the "Landscape, Environment, and Ecology" track was the most enduring of the five project tracks. This statement is based on the number of community initiatives inspired by the landscape framing during the course of the USUDS project (2012–2014) and those enabled by the project strategic vision after the project ended. Additionally, landscape strategies proposed by the project were the only USUDS strategies to receive technical support for development and more recently being considered for funding for implementation after the project ended.

9 For a detailed description of the Saida Green-Blue Network refer to Saida USUDS Green Open Spaces [23].

10 Al-Sabbagh builds on the Saida USUDS to explore the potential of an ecological landscape planning methodology to propose context-responsive and culture-inclusive planning tools to counter market-led, homogenizing, and exclusive development [24].

Responding to the second question, 'why was the landscape component more enduring', is more difficult. The reasons lie partly in the landscape's expansive framing of the city and its potential to integrate nature and people, environmental health, and heritage by contextualizing development strategies. The pace- and culture-responsive landscape strategies proposed were meaningful to the urban inhabitants at a time when 'urban infrastructure' projects required considerable technical and financial resources to implement and the 'Heritage' track was, for the most part, confined to the historic core. 'Economic betterment' and 'Governance' strategies, on the other hand, are long-term and not easily communicated to the public. In contrast, the Green-Blue Network served as the spatial embodiment of urban nature that is valued by all[11]. Similarly, the focus on rivers, streams and the coast resonate with the local discourse on Saida's outstanding natural setting and are integral to the shared memory of the city inhabitants, poor and rich, Lebanese and Palestinian.

On a practical level, the discursive elasticity of the landscape, enabled the landscape consultant to engage local players, individuals, and institutions, who were asked to contribute in various ways to the Green-Blue Network. Contributions by three of these players is herein discussed. First, 'Shajar w Bashar' is an NGO founded in 2008 in Saida committed to socially-disadvantaged youth and women. Shajar w Bashar programs aim to build capacity through training in decision-making skills that enhance social responsibilities and help improve living conditions. The NGO was identified as a key actor, and was approached to work on the abandoned railway corridor seeing as it had already engaged youth in a clean-up campaign for parts of the railway corridor. The USUDS team connected 'Shajar w Bashar' with Saida Municipality, experts and NGOs[12] that worked on similar projects to help build a strategy and portfolio for the project. The multifaceted framing of the abandoned railway corridor, and the fact that it was part of a city-wide strategic development, enabled Shajar w Bashar to secure funding for the project from PACE/USAID. The project "Mamar[13] 2013/2014" advocated and lobbied with local associations and youth from all religions, Palestinian and Lebanese, to claim the abandoned railway as a car-free, amenity space[14] (Figure 5).

Figure 5. 'Shajar w Bashar' initiative building on the landscape concept of the USUDS landscape proposal for reviving the abandoned railway corridor as a city-scale, inclusive public space.

---

11 Saida USUDS working group meetings.

12 'Train/Train', https://www.facebook.com/traintrainlebanon/ a group aiming to reclaim the railway network in Lebanon as a public realm. The extensive railway network in Lebanon was abandoned in the 1970s during the civil war. The corridors and stations are state-owned, allowing for temporary amenity use. Another group was 'Dictaphone', civil activists with the aim of reclaiming the public realm; see http://www.dictaphonegroup.com/both accessed (1 January 2018).

13 In Arabic 'Mamar' implies passage, which has spatial dimensions and non-spatial connotations of a process that addresses issues of governance for Saida youth.

14 The activities organized for the project by 'Shajar w Bashar' include: meetings for civic empowerment, 2013; meetings with political stakeholders; launching the project, 28 March 2014; petition signing; 'Walk/Cycle the Railway', 12 April 2014; closing ceremony 26 June 2014. See https://www.peaceinsight.org/conflicts/lebanon/peacebuilding-organisations/shajar-w-bashar/ and https://arab.org/directory/shajar-w-bashar/.

Another offshoot from the initiative was by Lil Madina[15], a group of young, professional civil activists. As with the previous example, a member of the Lil Madina group was commissioned by the lead author to research and document the agriculture heritage of Saida and the historic irrigation network[16], 'Qanat Al-Khaskiyeh' or 'Qanaya', that dates to Roman times. The network includes cultural practices to ration water quantities from the Awali River equitably. Saida orchards are famed for the quality of their produce and valued by all as integral to the collective identity. Yet, the orchards and the historic Qanaya were threatened by market-led development targeting these lands for exclusive housing development. Incorporating the orchards and Qanaya network, rivers, and streams within the USUDS landscape concept, revived public interest and awareness of their historic and cultural value and with the forward-looking landscape strategy, their potential in a sustainable vision for future development of Saida. Lil Madina has taken up these sentiments and succeeded during, and after, the USUDS to mobilize public and local organizations to protect these valued landscapes as a natural and cultural heritage of Saida (Figure 6).

Figure 6. 'Lil Madina' website profiling the historic Qanaya and Saida orchards as a collective landscape heritage of Saida.

With the Saida USUDS City Development Plan formally approved, Lil Madina focused on two very important initiatives. The first, East Wastani Development, aimed to combat realty development destined to displace orchards and agricultural lands north of the city. The project has strong political backing and is favored by the municipality. From 2012, Lil Madina conducted workshops[17] to find sustainable and community-inclusive alternatives to the corporate, generic development proposed. The outcome was promising as the political parties involved conceded and commissioned a master plan that followed the USUDS vision and guidelines[18].

---

[15] Arabic 'Lil Madina' translates 'for the city', founded in 2013 by residents of Saida, mainly architects and urban designers. Their collective agenda, modus operandi, and outreach differs considerably from 'Shajar w Bashar'. See https://lilmadinainitiative.wordpress.com/ (1 February 2018).

[16] This article's author, Sabbagh, participated in data gathering and field survey for the Qanaya study and was an active partner in many of the Lil Madina initiatives.

[17] Workshops co-organized by Lilmadina and Al-Sabbagh include: "Development of East Wastani", February 2014; Public and focused talks, i.e., Makassed Talk, May 2015; 'Towards Improving the Land-Pooling and Subdivision Tool Roundtable in collaboration with Issam Fares Institute, 19 December 2015 co-organized with Sabbagh; "Developing Al-Qamleh Valley" April 2015.

[18] The Lebanese urban planning framework applied by corporate land developers includes a set of regulatory (e.g., master plans, building law) and operational tools (e.g., land pooling tool) that is top-heavy, ignoring the specificity of place and local community. For a detailed discussion of the land pooling corporate application in the East Wastani project versus ecological guidelines for planning see Salwa Sabbagh, 2015 [24].

The second initiative, a direct offshoot from the Saida USUDS, aimed to revive the ecological integrity of the Qamli stream, another component of the Green-Blue Network. Lil Madina mobilized neighborhood residents, technical expertise, and secured funding to clean the river and conceptualize it as an amenity corridor. Accepting that river corridors bridge administrative boundaries and that cleaning the watercourse should necessarily involve other municipalities upstream from Saida, Lil Madina worked in close collaboration with adjoining municipalities, members of the Saida-Zahrani Union of Municipalities, securing funding to clean the Qamli by remapping the cadastral boundaries of the corridor, noting building encroachment, and repairing sewage discharge network [25].

The repercussions of the landscape strategies proposed by the Saida USUDS was not limited to enabling and empowering local communities, but was equally empowering of the Saida municipality. In less than a year after completion of the project, the Municipality of Metropolitan Barcelona proposed a series of workshops with the Saida municipality to develop detailed designs and cost estimates for components of the Blue-Green Network. The designs were completed in 2015, and the municipality proceeded to solicit funds for implementation. With a clear vision and development strategy secure, the Saida municipality was among the Lebanese cities and towns shortlisted in 2017–2018 by the French Development Agency (AFD) and the European Investment Bank (EIB) for potential funding of municipal projects. Once more, components of the Green-Blue Network attracted potential funding, seeing that they combined environmental health, social benefits, and quality living. Thus, for the first time, the Saida municipality was taking the initiative independently from the central government, backed by a collective vision for the future of Saida conceived by the city inhabitants. These small steps have come to dispel the prevailing perception in Lebanon that reduces the role of municipal authorities to paving streets and collecting solid waste. Empowering local authorities and changing their modus operandi, however, requires more than one project. Rather it should be seen as a process with successes and failures. There is a need to build competencies and technical skills to realize the shift towards empowered and visionary local governance.

The discourse emerging from the Saida project is also significant in that it reaffirms the claim made earlier of the usefulness of the case study research method in landscape architecture. Not only does the case study allow for the sharing of participatory tools, but it provides innovative problem-solving skills for planning the public realm in Lebanon as well, for example, the Green-Blue Network. Case studies are also an effective method for teaching by example. Incrementally, case studies can contribute to "a body of criticism and critical theory" and can "disseminate the effectiveness of landscape architecture outside the profession" [17] (p. 9).

5. Conclusions

Decentralization is redefining the relationship between citizens and the state towards stronger, democratic local and regional governance. The speed and trajectories of state decentralization, however, varies from one place to another, which is especially true when comparing governance in countries of the global north with those of the global south. The challenges of state decentralization in oppressive political systems and/or those lacking political accountability is immense [26,27]. Lebanon is one case in point. Accepting that participatory planning has the potential to ease the way to state decentralization and to address political struggles and power relations, as demonstrated by the Saida case study, the process is slow and the transformation, at best, long term. Nevertheless, enabling local authorities and empowering the public to voice and claim their right to the city, is one way to social justice and democratic societies [28].

The Saida case study was also an opportunity to demonstrate the expansive landscape design approach through participatory, contextualized, and multifunctional framing. The latter is especially significant in the absence of planning and integrative policies in Lebanon. Urban infrastructure, social welfare, and environmental health are compartmentalized in planning policies, the responsibility of different ministries, respectively, the Ministry of Public Works and Transportation, the Ministry of Work and Social Affairs, and the Ministry of Environment. There is often little coordination of the overlapping responsibilities of those ministries. The holistic landscape framing, as such, can provide a platform for the collaboration of the various state bodies and agencies through projects that are led by local authorities.

Author Contributions: Jala Makhzoumi was commissioned by Saida Municipality as national consultant for the "Landscape, Environment and Ecology"; Salwa al-Sabbagh contributed to the Saida USUDS data collection, field work and participated in meetings as part of her the Master of Urban Design thesis, American University of Beirut.

Conflicts of Interest: The authors declare no conflict of interest.

References

1. Oxhorn, P. Civil society without a State? Transnational civil society and the challenge of democracy in a globalizing world. *World Future* **2007**, *63*, 324–339. [CrossRef]
2. Oxhorn, P.; Tulchin, J.; Selee, A. *Decentralization, Democratic Governance, and Civil Society in Comparative Perspective. Africa, Asia and Latin America*; John Hopkins University Press: Baltimore, MD, USA, 2004.
3. Prachett, L.; Wilson, D. Local Government under Siege. In *Local Democracy and Local Government*; Prachett, L., Wilson, D., Eds.; Macmillan Press Ltd.: London, UK, 1996; pp. 1–19.
4. Pierre, J. Public-Private Partnerships and Urban Governance: Introduction. In *Decentralization, Democratic Governance, and Civil Society in Comparative Perspective Africa, Asia and Latin America*; Oxhorn, P., Tulchin, J., Selee, A., Eds.; John Hopkins University Press: Baltimore, MD, USA, 1997; pp. 1–10.
5. Stoker, G. Public-Private Partnerships and Urban Governance. In *Urban Governance*; Pierre, J., Ed.; Partnerships, Palgrave Macmillan: London, UK, 1998; pp. 34–51.
6. UNHabitat, Governance. Available online: https://unhabitat.org/governance/ (accessed on 27 September 2017).
7. McEwan, C. Bringing government to the people: Women, local governance and community participation in South Africa. *Geoforum* **2003**, *34*, 469–481. [CrossRef]
8. Williams, G. Evaluating participatory development: Tyranny, power and (re)politicization. *Third World Q.* **2004**, *25*, 557–579. [CrossRef]
9. Participation Research Cluster. Participatory Methods. Available online: http://www.participatorymethods.org/ (accessed on 27 September 2017).
10. Council for Development and Reconstruction, National Physical Master Plan of the Lebanese Territories. 2005. Available online: http://www.cdr.gov.lb/study/sdatl/sdatle.htm (accessed on 1 January 2018).
11. Makhzoumi, J. The greening discourse: Ecological landscape design and city regions in the Mashreq. In *Reconceptualizing Boundaries: Urban Design in the Arab World*; Saliba, R., Ed.; Ashgate: London, UK, 2015; pp. 63–80.
12. Egoz, S.; Makhzoumi, J.; Pungetti, G. *The Right to Landscape: Contesting Landscape and Human Rights*; Ashgate: London, UK, 2011.
13. Olwig, K. *Landscape, Nature and the Body Politic*; University of Wisconsin Press: Madison, WI, USA, 2002.
14. Egoz, S.; Jorgensen, K.; Ruggeri, D. *Defining Landscape Democracy*; Edward Elgar Publishing: London, UK, 2018.
15. Wall, E.; Waterman, T. *Landscape and Agency: Critical Essays*; Routledge: London, UK, 2018.
16. Gillham, B. *Case Study Research Methods*; Bloomsbury Publishing: London, UK, 2010.
17. Francis, M. *A Case Study Method for Landscape Architecture*; Landscape Architecture Foundation: Washington, DC, USA, 1999. Available online: https://lafoundation.org/myos/my-uploads/2010/08/19/casestudymethod.pdf (accessed on 12 February 2018).
18. Makhzoumi, J. Landscape in the Middle East: An inquiry. *Landsc. Res.* **2002**, *27*, 213–228. [CrossRef]
19. Chaaban, J.; Srour, I.; Hamadeh, K. Saida USUDS Strategic Diagnosis: Institutional and Legal Framework. 2014. Available online: http://www.medcities.org/documents/22116/135803/5.+Saida_Diagnosis+_+Institutional+and+Legal++Framework.pdf/b62f3b3a-53aa-41d8-bdbf-4fccbe2d4114 (accessed on 2 January 2018).
20. MedCities. Available online: http://www.medcities.org/ (accessed on 31 December 2017).
21. MedCities Saida CDS. Available online: http://www.medcities.org/web/saida (accessed on 2 January 2018).
22. Makhzoumi, J.; Pungetti, G. *Ecological Design and Planning: The Mediterranean Context*; E. & F. N. Spon, Routledge: London, UK, 1999.

23. Makhzoumi, J.; Al-Sabbagh, S. Saida USUDS Strategic Diagnosis: Green Open Spaces. 2014. Available online: http://www.medcities.org/documents/22116/135803/2.+Saida_Diagnosis_Green-Open.pdf/4bbf1534-8be5-46b1-b2ea-1b30f477d27d (accessed on 2 January 2018).
24. Al-Sabbagh, S. Rethinking Planning Tools through the Ecological Landscape Design Approach: Saida Case Study. Master's Thesis, Urban Design, American University of Beirut, Beirut, Lebanon, February 2015.
25. Lil, M. *Qamleh River in Greater Saida: Its Disappearance and Efforts to Revive it*; Creative Commons: Saida, Lebanon, 2017. (In Arabic)
26. De Soto, H. *The Other Path, The Economic Answer to Terrorism*; Basic Books: New York, NY, USA, 2002.
27. De Soto, H. *The Mystery of Capital: Why Capitalism Triumphs in the West and Fails Everywhere Else*; Basic Books: New York, NY, USA, 2000.
28. Makhzoumi, J. Landscape architecture and the discourse on democracy in the Middle East. In *Defining Landscape Democracy*; Egoz, S., Jorgensen, K., Ruggeri, D., Eds.; Edward Elgar Publishing: London, UK, 2018.

land

Article

Landscape Risk Assessment Model and Decision Support System for the Protection of the Natural and Cultural Heritage in the Eastern Mediterranean Area

Maria Gabriella Trovato *, Dana Ali, Jessica Nicolas, Ammar El Halabi † and Sarah Meouche ‡

Landscape Design and Ecosystem Management, Faculty of Agricultural and Food Sciences, American University of Beirut, Riad El-Solh, Beirut 1107-2020, Lebanon; dra10@mail.aub.edu (D.A.); jessicatnicolas@gmail.com (J.N.); aae48@mail.aub.edu (A.E.H.); sarah.meoushe@gmail.com (S.M.)

* Correspondence: mt63@aub.edu.lb; Tel.: +961-78827535

† Current Affiliation: Freelance Software Developer, Beirut, Lebanon.

‡ Current Affiliation: Freelance Landscape Architect, Beirut, Lebanon.

Received: 18 July 2017; Accepted: 21 October 2017; Published: 3 November 2017

Abstract: In recent years, the competition of uses for scarce and highly valuable natural resources, and the frequency and severity of natural and technological disasters have increased, and this trend is likely to worsen in the years to come. In the Mediterranean area, especially in its Eastern part, the high human exploitation driven by different economic sectors and interests is resulting in intensive use of the land and its resources. Tourism intensification, rapid growth of urban settlement and related sprawl, movement and displacement of populations, rural abandonment, and adoption of different agricultural techniques are profoundly and rapidly changing the landscape character of the East Mediterranean. In view of the risks to cultural and natural heritage, a Landscape Risk Assessment Model (LRA) and Decision Support System (LDSS) were developed through the MedScapes-ENPI project. This paper reports the experience conducted at the Landscape Design and Ecosystem Management Department (LDEM) in the American University of Beirut (AUB) in developing the two tools, LRA and LDSS. It aims to provide insight into the methodology designed and tested during the length of the project to take into account the protection of landscapes of particular interest as well as the rational planning of all the landscapes with special emphasis on the use of natural resources. The assessment was applied in the study area of each partner country of the ENPI project, allowing for a better understanding of the implications in land-use and conservation decision-making.

Keywords: Landscape Risk Assessment Model; Landscape Decision Support System; East Med landscape

1. Introduction

In the Mediterranean area, especially in its Eastern part, the high human exploitation driven by different economic sectors and interests is resulting in intensive use of the land and its resources. The consequences are evident and the fact is that never before have we transformed so much territory as in recent years and at such a fast pace [1]. The MedScapes project, through the development of a methodology to assess and evaluate the landscape of the East Mediterranean and its threats (Landscape Character Assessment (LCA), Landscape Risk Assessment (LRA), Landscape Decision Support System (LDSS)) aimed to introduce the concept of landscape planning as a tool to integrate the different features of the region (the ecological and cultural ones) to support landscape strategies and planning. Working on the implementation of the European Landscape Convention (ELC, 2000 [2]), it intended to support local authorities and non-governmental organization with the development of scientific tools designed for the protection, conservation, and management of the natural and cultural heritage of the region.

The objective of the present study was to asses, using GIS, the loss of the landscape value of the chosen areas in the four partner countries under the pressure of selected and measured hazards, and to develop a landscape spatial decision support system to support decision-making, and assist stakeholders in comparative assessment and selection of options for change. The review of already implemented and tested Risk Assessment (RA) and Decision Support System (DSS) approaches revealed that most of them are models for ecological analysis and assessment (i.e., EcoRa, Ecological risk assessment [3], DS WMVOC Environmental Risk Assessment for Decision Support System for water management in the vicinity of open cast mines [4]). Some of them are directed to produce integrated multi-risk maps to achieve more effective spatial planning procedures in areas prone to natural disasters (i.e., ARMONIA, 2007 [5]), to encourage climate adaptation planning in coastal areas (i.e., DESYCO, 2015 [6]), or to intervene at a regional scale to support environmental risk assessment and management (i.e., SYRIADE, 2012 Spatial decision support system for Regional risk Assessment of Degraded land [7]).

Despite the fact that the knowledge, recognition, and awareness of a landscape as a value have become a very distinctive topic in the last decades, there is still a deficiency of tools specifically designed for the landscape issue. Research studies have been conducted on the assessment of the landscape value [8] and have been motivated by the need to inform and enhance land use planning and environmental management. For example, typologies of landscape values have been developed and implemented to inform forest management [9–11], national parks and protected area management [12], urban park planning [13], residential and tourism development [14,15], coastal area management [16], rural development [17,18], and climate change risk [19]. Based on the consideration that the value of a landscape is an important issue for landscape protection planning [20] and that it is a key problem pertaining to sustainable landscape management [21–23], we concentrated our study on the assessment of the loss and depletion of landscape value in order to be able to produce scenarios for more coordinated planning.

This paper describes the methodology we developed and implemented to analyze the landscape hazards, landscape values, and landscape risk and to build an interactive web-based interface, where the users could view the results of the Landscape Character Assessment (LCA) and LRA and alter scenarios in the LRA pertaining to intensification of urban sprawl patterns hazard and Landscape Value (LV). This method was built upon the results of the LCA conducted in the four Eastern Mediterranean partner countries of the project (Cyprus, Greece, Jordan, and Lebanon), and takes into account two main components: the landscape value, as a result of natural and cultural elements; and the severity of risks which cause a threat/pressure for change. Four hazard categories were chosen and assessed: desertification, intensification of urban sprawl patterns, erosion, and forest fire.

In this study, the landscape value is quantitatively evaluated by using three different parameters (Dominant habitat, Naturalness, and Habitat Continuity), while the hazards are quantitatively and qualitatively measured. The risk is assessed with reference to the potential occurrence of the previously mentioned hazards. Landscape hazard frequency may impoverish the value of the landscape, directly depleting natural resources and features (trees, woods, streams), altering ecological functions by landscape fragmentation, and eroding high quality agricultural land [24].

This study presents some limitations due to the complexity in comparing and interrelating data from four different countries, the difficulty in acquiring the same type of materials in the four partner countries, and the impossibility of monitoring the changes through time due to the lack of data. This paper acknowledges the 'profound transition' that Antrop [25] asserts the landscape concept is passing through and it is visible on the indicators chosen and the weight we allocate to them.

2. Materials and Methods

2.1. Study Area and Available Data

The areas investigated by the RA and LDSS were the Level 1 LCA defined zones in the four partner countries of the project (Figures 1 and 2).

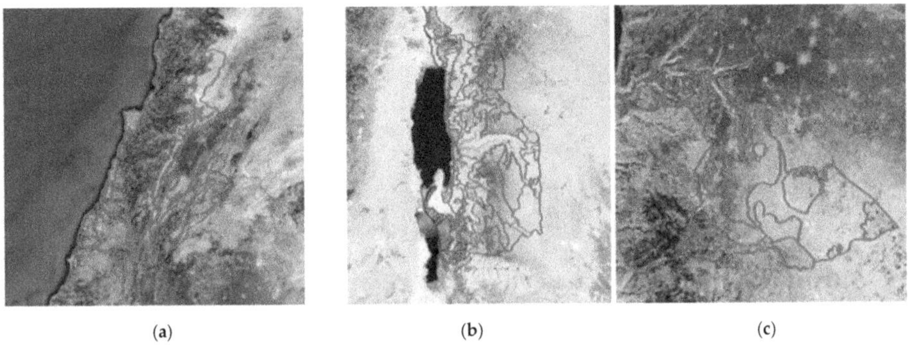

Figure 1. Risk assessment model study areas in the partner countries. (a) Lebanon Area: 4747 km^2; (b) Jordan—Mujib Area: 6790 km^2; (c) Jordan—Al-Yarmuk Area: 3800 km^2.

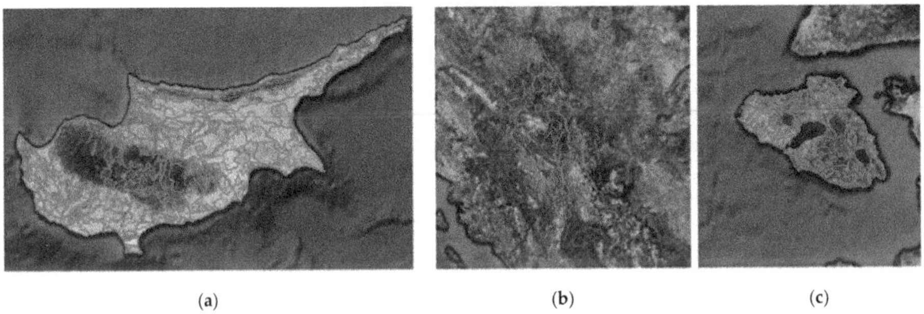

Figure 2. Risk assessment model study areas in the partner countries. (a) Cyprus Area: 9251 km^2; (b) Greece—Epirus Area: 6790 km^2; (c) Greece—Lesvos Area: 1632 km^2.

The study area in Lebanon represents 45% of the country and it encompasses four distinct geomorphological regions: the coastal plain, Lebanon Mountain range, Bekaa Valley, and the Anti-Lebanon mountain range.

In Jordan, two areas were chosen: the first, Mujib, is located within Madaba and Kerak governorates and it extends from the Jordan Rift Valley (JRV) escarpment in the west, at 420 m below sea level, to the central highlands plateau in the east, with elevations exceeding 700 m above sea level; the second, Al-Yarmouk catchment area, is located in the North-West part of Jordan and it is comprised of large urban centers (Irbid and Al Ramtha) in addition to extensive agricultural activities, and it includes unique heritage and historical places of Jordan.

Two areas were also selected in Greece: Epirus, located in the Region of Epirus, predominantly mountainous and characterized by significant landscape diversity (high mountains, cultivated settled areas, and grazed pastures); and Lesvos, the third largest island in the Aegean Sea, with approximately 41% of the area planted with olive, 34% with maquis and garrigue, 17% with forest, and 3% with other crops, while the remaining land has various uses such as constructions, wetlands, etc.

As for Cyprus, the entire island was part of the study. The island is divided into three geomorphological zones: the Troodos Mountain, the Pentadaktylos Range, and the Mesaoria plain.

The six areas in the four countries present differences in geomorphology, land use, settlement patterns, and economy. This diversity had to be taken into consideration while developing our tools and constructing a conceptual model holistically.

2.2. Methodology

The methodology was the result of an extensive analysis and investigation throughout prevailing studies and RA-DSS models developed and implemented in past years. Disaster management became of particular importance in recent years [26] and disaster risk reduction is a world challenge. The 1992 Earth Summit on Environment and Development in Rio de Janeiro, Brazil, provided a venue for international recognition of the connection between development, environment, and disaster risk reduction [27]. Thereafter, passing from the 'environmental justice' perspective and the Environmental Protection Agency, EPA, the international research evolved and moved towards considering the cumulative impacts of multiple hazards. In Europe, the European Commission has been encouraging studies on natural disasters since the early 1980s. Research under the Seventh Framework Program aimed to reduce and mitigate the environmental, social, and economic effects of natural disasters through a holistic multidisciplinary approach in which aspects of hazards/multi-hazards, vulnerability, and risk assessment were addressed in an integrated manner [28].

The literature on sustainability and hazard or risk mitigation is now extensive and includes academic, practitioner, and governmental consideration of the need to adopt sustainable approaches to hazard and risk management [29]. While environmental risk assessment is broadly applied to identify the danger the environment is exposed to in relation to the different types of hazards, on the other hand, landscape risk assessment related to the landscape value is an under-evaluated and investigated tool. Landscape risk appears to be a more uncertain issue where approaches are focused on the visual impact of changes in land use due to human activities (i.e., the construction of new buildings or infrastructures) [30]. Overall, very few publications connect the assessment of landscape with land use decision-making. For example, MULBO model framework (Multi-criteria Landscape Assessment and Optimization) is a spatially explicit decision support method on the basis of risk evaluations for landscape function that aims to establish optimal land use patterns as scenarios that are balanced compromises between conflicting goals for the reduction of assessed risks [21].

As a result of the ELC, the Discrete Choice Experiment (DCE) method was adopted to enrich the process of landscape decision-making mainly due to its flexibility and ability to take into account the multidimensional nature of landscapes and to provide detailed information about marginal changes in landscapes as well as trade-offs between the landscape attributes themselves and between the landscape attributes and money [31]. The DCE seems to better respond to the need of experts and policy-makers in finding a way to quantify the value of landscapes to the local residents and to incorporate these values into their landscape protection, management, and planning decisions [32]. On the other hand, the DCE does not take into account and does not assess the landscape changes in relation to the different types of hazards.

The LRA-LDSS framework, presented in this paper, incorporates some important key aspects:

- The LRA was conducted at the scale of Landscape Descriptive Units (LDUs) that were delineated by the LCA process. This allowed the tool to visualize the hazard impact for the different Landscape Character Types.
- The LRA was formulated through a risk equation as the result of the relation between the two factors Hazard and Landscape Value
- The framework made use of a two-dimensional matrix for a risk legend, borrowed from the ARMONIA EU project, that was reworked and adapted to our study.

- The LRA-LDSS used the scenario-based approach to spatially visualize a series of possible physical landscape transformations due to the increased intensity of a landscape hazard.

Our conceptual framework is the result of the relation between LCA, LRA, and LDSS and describes the operative steps used to derive hazards maps, landscape values maps, risk maps, and scenarios' maps (Figure 3). It is organized as per the following scheme:

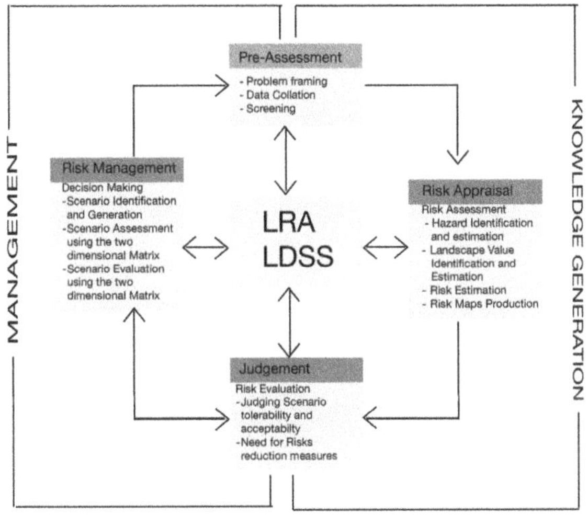

Figure 3. Landscape Risk Assessment conceptual framework.

3. Landscape Risk Assessment (LRA) and Landscape Decision Support System (LDSS)

The Landscape Risk Assessment was conducted in phases: Pre-Assessment, Risk Appraisal, Judgment/Risk Evaluation, and Risk Management.

The Pre-Assessment allowed us to frame the aims of the model, to verify the availability of data per countries, to collate the LCA results, and to test and assess the correct model in line with the objective of the project.

The Risk Appraisal was based on hazard identification, hazard ranking and assessment of magnitude per Landscape Descriptive Units (LDUs), Landscape Values identification, and assessment per each of the LDUs.

The Judgment/Risk Evaluation was conducted by determining the significance of the estimated risks in relation to the loss of the assessed Landscape Values. It was carried out by applying the MedScapes risk equation: Risk = Hazard × Landscape Value derived from the original equation that we borrowed from the EU project ARMONIA [5].

The Risk Management phase let the user to build scenarios through the LDSS Web user's interface. The Landscape DSS support future land decisions in relation to the protection of the natural and cultural heritage of the Eastern Mediterranean area.

3.1. Pre-Assessment Phase

Building on the results from the Landscape Character Assessment was a key point in our strategy, which integrates results of level 1 and level 2 LCA assessments in the hazard analysis and landscape value analysis, respectively.

Prior to reaching the final model an assessment of the availability of material was conducted between the partners, discussions took place, and questionnaires were handed out. Partners were asked to share their opinions and expertise on the most pressing threats to the character of the landscape in the East Mediterranean region and on the landscape value appraisal methods. This formulated the basis on which the conceptual model was developed. Data collation then followed and continued in parallel with the development and refinement of the Risk Assessment model. The process included a back and forth interaction between the partners and the Lebanese team, through Skype and email, where discussions on data availability, coherence, and verification took place in order to achieve a comprehensive and homogeneous methodology. Limitations of available data led us to a further refinement of the model to suit the situations observed in the four partner countries.

3.2. *Risk Appraisal Phase*

3.2.1. Landscape Hazard Identification

Landscape hazard appears to be a more uncertain and less explored issue, where approaches have focused on the visual impact of changes in land-use due to human activities, such as the construction of new buildings or infrastructure [33,34]. Antrop [35] has focused on four different categories of pressures on landscapes that can be strictly interrelated: housing and settlements, land as production space, infrastructure, and the use of land as recreation and fragmentation. In the MedScapes work, hazards were defined as potential threats that are likely to cause damage to the landscape character in the event that they occur [24]. The landscape hazards were chosen according to the following criterion: severity in terms of the effect of the occurrence at a location (site parameter), applicability across the partner countries, availability of information, and relevance to results from the Landscape Character Assessment (LCA). Research on the most pressing threats to the Mediterranean countries, data entry on the most pressing threats per LDU, and discussions with the partners of the MedScapes project yielded the following list of hazards to be assessed: Desertification, Erosion, Forest Fires, and Intensification of Urban Sprawl Patterns. The partners ranked the hazards as per the following Table 1. Hazards ranking table, in which the value 6 represents the bigger perceived threat and 0 being the least.

Table 1. Hazards ranking table.

Hazard	Jordan Yarmuk	Jordan Mujib	Cyprus	Greece Lesvos	Greece Epirus	Lebanon	Total
Intensification of urban sprawl patterns	6	0	5	6	5	6	28
Erosion	5	0	3	5	6	5	24
Desertification	4	4	4	4	4	4	24
Loss of agricultural lands	3	5	2	0	2	3	15
Forest fires	1	0	1	3	1	2	8
Landscape transformation by new planned developments	2	6	6	0	3	1	18
Flood	0	0	0	0	0	0	0

Based on LCA data, and research on the most pressing threats on the Eastern Mediterranean landscape, four types of quasi-natural and man-made hazards where chosen:

- Desertification (qualitatively assessed)
- Erosion, (qualitatively assessed)
- Forest Fires (qualitatively assessed), and
- Intensification of urban sprawl patterns (quantitatively assessed)

3.2.2. Landscape Hazards Ranking and Assessment

Hazard assessment is the key element of risk analysis. It is carried out by processing info taken from different kinds of maps and/or field-work. In our research the evaluation was conducted at the level of the LDUs and the hazard rating was carried out using the related GIS maps (Erosion, Forest fire,

Desertification, and Intensification of urban sprawl) and data provided by the project's partners based on their availability. The magnitude of the risk per each of the LDUs was identified and ranked using the qualitative classification with a score ranking from 1 (minimum hazard) to 4 (maximum hazard) as per Figure 4.

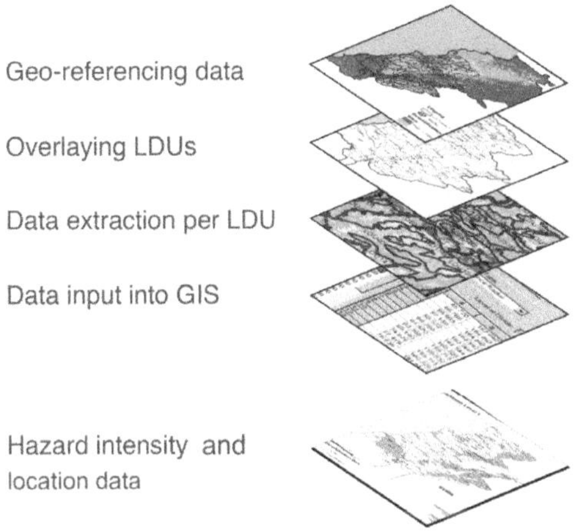

Value ranges used unanimously:

Rank	Value	Value Range
1	Low	0–25
2	Low-Medium	25.1–50
3	Medium-High	50.1–75
4	High	75.1–100

Figure 4. Hazard assessment methodology. LDU, landscape descriptive unit.

The Intensification of Urban Sprawl Patterns was the only hazard that we were able to quantitatively assess using four parameters: existing settlement, landform, geology/slope, and informal settlements. The above were selected based on a study by La Rosa and Martinico in Sicily, Italy [24]. The criteria chosen to weight the parameters emphasized the importance of existing settlement patterns and geology/slope, with weights of 40% and 30%, respectively. We assigned lower weights for landform (20%) and informal settlements (10%) since their assessment is based on the existing trends of settlement/landform and the large influx of refugees. The data used includes: the spatial data from the Landscape Character Assessment, which was indexed based on current trends and on relationships between attributes; as well as the United Nations High Commissioner for Refugees (UNHCR) map of Syrian refugee camp locations, which were considered for the assessment of the impact of informal settlements on intensification patterns. In Lebanon, the large influx of refugees has played a role in altering the landscape in key areas such as the Bekaa Valley, Coastal South Lebanon, and North Lebanon. The assessment and ranking methodology is shown in Table 2.

Table 2. Intensification of Urban sprawl: Assessment and ranking.

Attribute											Formula	Total Value
LANDFORM	water body	deep valley	mountainous areas	valley	high uplands	uplands	high plateau	hills	plateau	Low lands	100/6 X RANK	20%
RANK	0	0	1	2	2	3	4	4	5	6		
SETTLEMENT	Unsettled	Sparsely Settled	Settled-Medium Spaced	Settled-Closely Spaced	Settled	Settled to Urban	Urban				100/4 X RANK	40%
RANK	1	2	3	4	0	0	0					
INFORMAL SETTLEMENTS	Unsettled	Sparsely Settled	Settled-Medium Spaced	Settled-Closely Spaced	Settled	Settled to Urban	Urban				100/4 X RANK	10%
RANK	4	3	2	1	0	0	0					
GEOLOGY/SLOPE	No Risk	Low	Medium	High								
RANK	3	2	1	0								
GEOLOGY	Limestone	Sandstone	Chalk	Bare Rock	Alluvials	Marl	Volcanics	Shales	Fluvials		100/3 X RANK	30%
RANK	1	2	3	4	5	6	7	8	9			
SLOPE % Range	0	1–12	13–20	21–35	36–50	51+						
RANK	6	5	4	3	2	1						

3.2.3. Landscape Value Identification

The authors of this study distinguish between landscape value and landscape character. Landscape character is concerned with the fabric that makes up the landscape of a place, usually in the absence of qualitative rating system, but also with how that landscape is perceived (e.g., the pattern of available viewing, Williamson, 1979 [36]). Value, on the other hand, is concerned with the quality of experience of place, associated in the literature with as diverse themes as visual variety (Leonard and Hammond, 1984 [37]), positive emotions (Daniel and Vining, 1983 [38]), and length of association with a landscape (RFA, 2002 [39]).

Values in landscape have been studied by numerous researchers in a variety of contexts [40]: Greg Brown and colleagues have, for example, undertaken extensive work in defining landscape values for participatory mapping [41–43]; Michael Jones has developed a categorization of values for addressing conflicts in planning issues [44]; Janet Stephenson has exposed how different forms of assessments take up certain landscape values whilst ignoring others [45]; while others have taken up the question of how professional values differ from those of laypersons professional contra lay person values [46]. Butler (2016), in particular, argues on how landscape character assessment represents a critical point in the framing of landscape values [40].

In our study, the landscape value was based on its physical character assessed throughout the LCA method implemented in the first phase of the MedScapes project considering the ecological and cultural qualities. To reach a comprehensive model that builds up on the LCA, integrates contemporary methods of evaluating the value of the landscape, and takes into account the expertise of the partner countries involved (Table 3). Partners' input), the following parameters were to be considered in the Landscape Value assessment:

a. Ecological Values (quantitatively assessed)
    - Naturalness of habitat
    - Habitat continuity
    - Dominant habitat type
b. Historical/Cultural Values (quantitatively assessed)
    - Archeological sites
    - Heritage/historical entities, trails, centers and museums

Table 3. Partners' input.

Greece	Cultural values (architecture, traditional settlements, field patterns, terraces) Environmental values (vegetation, geomorphology, fauna) Historical values (archeological and monumental sites) Economic values (agricultural, tourist)
Cyprus	Cultural Integrity (change of use, survival of cultural pattern, visual impact of change) Ecological integrity (naturalness of LDU, habitat continuity, number of main habitat types, dominant type, intensity of management)
Jordan	Geology Natural Landforms Vegetation Fauna environment Landscape diversity Historical roads and town centers

3.2.4. Landscape Value Assessment

To assess the value of the landscape we analyzed and assessed data extracted from the LCA process, land cover and land-use maps, satellite imagery (which was an asset, considering the scale of the study), and historical/cultural maps produced by governmental institutions. The two main parameters were weighed at 80% (ecological integrity) and 20% (historical/cultural values).

Ecological Integrity

Adhering to the LCA standards of the MedScapes project, Ecological Integrity was assessed based on the following components: habitat continuity, dominant habitat, and naturalness of LDU (Table 4).

Table 4. Ecological Integrity ranking method.

						Value	Total Value per LDU/100
Habitat Continuity	Continuous	Linked Patches	Separate Patches	Fragmented		/4	value/4
Rank	4	3	2	1			
							+
Habitat Type	Herbaceous	Low Scrub	Tall Scrub	Woodland	Cultivation		
Rank	1	2	3	4	2	/4	Value/4
							+
Naturalness of LDU	Mostly natural	Semi Natural	Even Mix	Mostly cultivated			
Rank	4	1	3	2		/4	Value/4
							=
						Total/12	**Cumm. Total/12 × 100**

Habitat Continuity (33.33%)

This assessment was carried out per LDUs by relying on high-resolution satellite imageries. A continuous landscape (rank 4) was the most valuable type, as characterized by being an uninterrupted patch with few construction/road networks, quarries, etc.; whilst a fragmented landscape (rank 1) was the least valuable type, marked by heavy urbanization and land-reshaping activities (Figure 5).

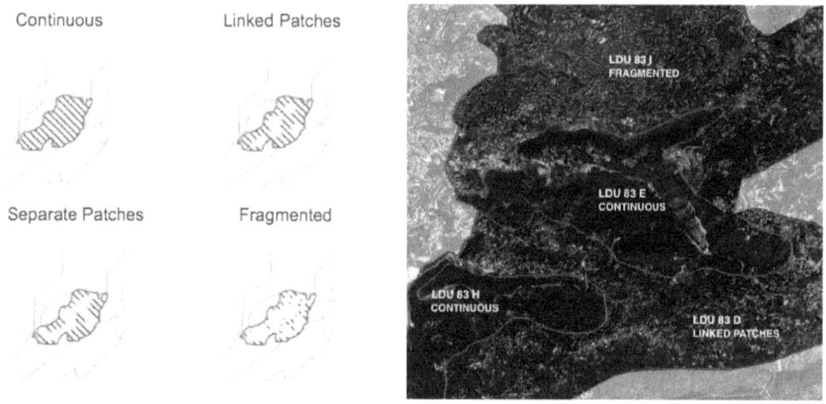

Figure 5. Habitat Continuity assessment per LDUs.

Dominant Habitat (33.33%)

The data were derived from land cover and land-use maps, taking into consideration the most dominant habitat type per LDU. Table 5 below reports the ranking system of each type and related description.

Table 5. Dominant habitat ranking system of each landscape type.

Dominant Habitat	Type	Rank	Description
Herbaceous	<0.2 m	1	Natural grasslands, rocky landscapes, bare rocks.
Low Scrub	0.2 to 1 m	2	Sclerophyllus beg.
Cultivation	Field crops, orchards, groves	2	Groves, agriculture, cultivation patterns.
Tall Scrub	0.5 m to 1.5 m	3	Transitional woodland-shrub.
Woodland	1.2 m	4	Forests (coniferous, broad leaved).

Naturalness (33.33%)

LDUs were categorized according to their naturalness, as shown in Table 6 below. As most landscapes had a mosaic of land-use types, this classification differentiates between landscapes that are minimally altered (mostly natural, rank 4), and landscapes that are mostly built up with still existing cultivation practices (semi-natural, rank 1) (Figures 6 and 7).

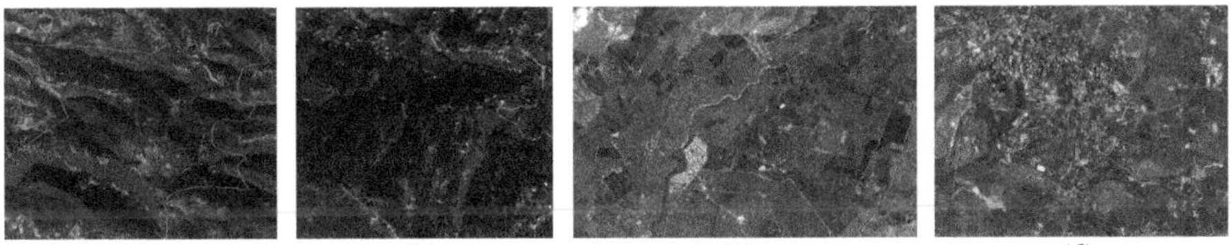

(a) (b) (c) (d)

Figure 6. Naturalness rank per each of the LDUs: (**a**) Mostly Natural, (**b**) Even Mix, (**c**) Mostly Cultivated and (**d**) Semi-Natural.

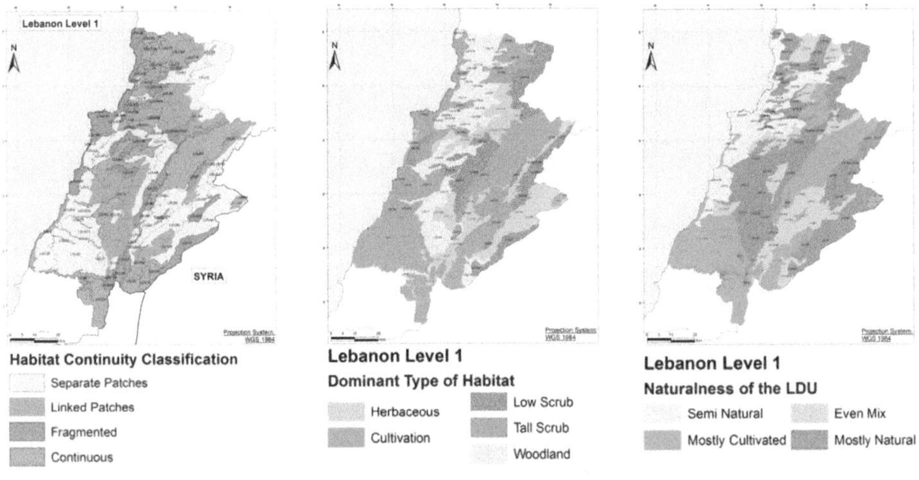

(a) (b) (c)

Figure 7. Maps of Lebanon's habitat continuity (**a**), dominant habitat (**b**), and naturalness (**c**).

Table 6. Ranking system of LDUs according to naturalness.

Naturalness	Rank	Description
Mostly Natural	4	A mostly natural LDU
Even Mix	3	A mix between a natural and cultivated LDU
Mostly Cultivated	2	Mostly agricultural field crops, orchards, groves, terraces
Semi Natural	1	A mix of unnatural and cultivated LDUs

3.2.5. Historical/Cultural Values

Cultural identity is strongly associated with the ways in which people interact with their landscapes. Much has been written about the significance of landscape (or the related idea of place) to communities and their cultural identity [47]. The literature ranges from sociological and anthropological work, to studies of place identities [48,49]. In our study, cultural values are assessed using the national maps and by ranking the data extracted from them based on the importance of the site/monument, on the dimension of the area/architecture, and on the regional and national significance. Two components were taken into consideration: Archeological sites, and Heritage/historical entities (trails, centers, museums, etc.). The assessment was quantitatively conducted considering the availability of spatial data.

3.3. *The Judgment/Risk Evaluation Phase*

The Risk Evaluation was conducted by determining the significance of the estimated risks in relation to the loss of the assessed Landscape Values by applying the MEDSCAPES risk equation:

$$\text{Risk} = \text{Hazard} \times \text{Landscape Value} \quad (1)$$

which was derived from the original equation we borrow from the EU project ARMONIA [21]:

$$\text{Risk} = \text{Hazard} \times \text{Expected losses} \quad (2)$$

where Expected losses is Vulnerability per Value.

In our approach, expressed by the above equation, Hazard is a potential threat that is likely to cause damage to the character of the landscape in cases where it occurs, and Landscape Value (LV) is the value of the assets in each Landscape Descriptive Unit (LDU), mainly being ecological, natural, and historical/archeological assets. The equation was portrayed through a two-dimensional matrix that we drew based on a study Greiving et al. performed on Integrated Risk Assessment of Spatially Relevant Hazards [50]. The same authors, having in mind the cost–benefit analysis in relation to the economic and physical risk, developed a similar matrix during the ARMONIA project. In the 4×4 matrix the hazard intensity and the LV degree were summed up to yield the integrated risk value. It was represented through a color code that identifies the risk classes from very low risk to very high risk (Figure 8). This tool also acts as a legend for the produced Landscape Risk Maps by decoding the value of the risks in each LDU as portrayed in the maps.

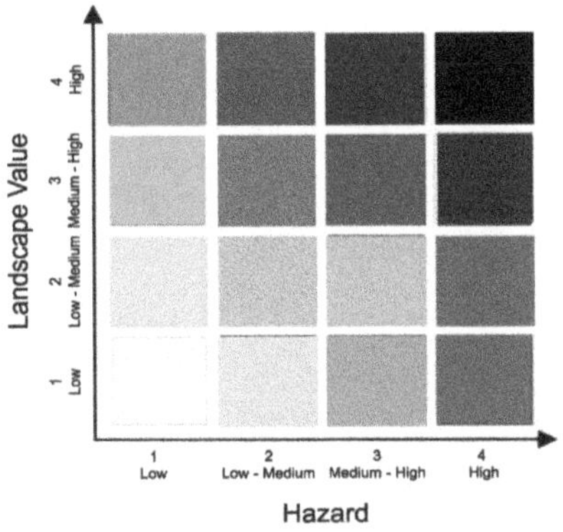

(a)

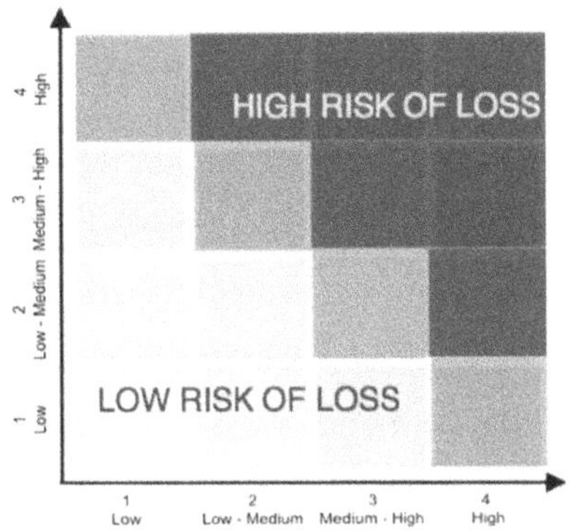

(b)

Figure 8. MEDSCAPES risk assessment (LRA) matrix: (**a**): Color code legend of the LRA; (**b**): Dark colors = High risk of loss, Light colors = low risk of loss.

3.4. The Risk Management Phase: Landscape Decision Support System (LDSS)

The Decision Support System is defined as a computer aided system to support decision-making, assisting stakeholders in comparative assessment and selection of options for change. It aids in scenario alteration and generation to support future decisions. Spatial Decision Support Systems (SDSSs) for natural resources management are computer-based tools that tightly integrate decision theory and models with ecological models and GIS analyses and mapping [51]. The information provided by SDSSs gives decision makers increased ability to follow outcomes of interacting variables, improves the reproducibility of decisions, and documents the reason why a particular choice was made [21]. The use of SDSSs in land-use and urban planning is a common approach in Western countries [52], but its application in the spatial planning of the East Mediterranean region is still not valued.

Planning is a future oriented activity and the future being planned for is always more or less uncertain. Nevertheless, an image of what the future may look like in important aspects is needed in order to provide a context for the actions considered for inclusion in the plan. For planning purposes most of the examples of SDSSs have been calibrated for the regional scale in a very case-specific way for solving one, or only few, functional problem(s) [53]; for example, for coastal zone management [54] and for monitoring purposes [55]. Advances have been made in the use of spatial models to assess the impact of potential development [56].

The tool proposed in MedScapes offers an integrative framework in landscape risk assessment and scenario building for the support of land-use decisions. It is map-centered and scenarios are spatially assessed. The use of a LDSS aims at helping stakeholders to spatially visualize future possibilities and alternatives to support their decisions.

Generally, the decision-making problems can be categorized under certainty conditions or under uncertainty conditions, according to the amount and typology of knowledge available to the Decision Making (DM) [57]. Due to the limited information available in the four partner countries and the difficulties encountered in collecting, comparing, and estimating data from different resources the problem description and analysis were conducted under uncertainty conditions.

The Landscape Decision Support System in MedScapes was developed as an interactive web-based interface that serves to implement the LRA approach while managing the input data. It provides LCA, Hazards, Landscape Values, Risk, and Scenarios maps for the identification and prioritization of areas, and the basis for the definition of adaptation and planning strategies and measurers. Through the LDSS the user can visualize the results of the Landscape Character Assessment, the Landscape Risk Assessment process and create spatially-based scenarios by altering the hazard pertaining to intensification of urban sprawl patterns and the Landscape Value (Figure 9).

It has the following key features:

- It maps and visualizes information on up to four different hazards and the consequent losses on Landscape Value at the scale of the LCA level 1 assessment (1:250,000 scale).
- It enables different scenarios to be run which generate information about hazards and landscape loss for all the study areas in the four partners countries and for each of the LDUs, so that different options for mitigating risks or developing land can be compared
- It enables editing scenarios by altering different indexing weights and LDU ranks
- It provides a knowledge base on hazards

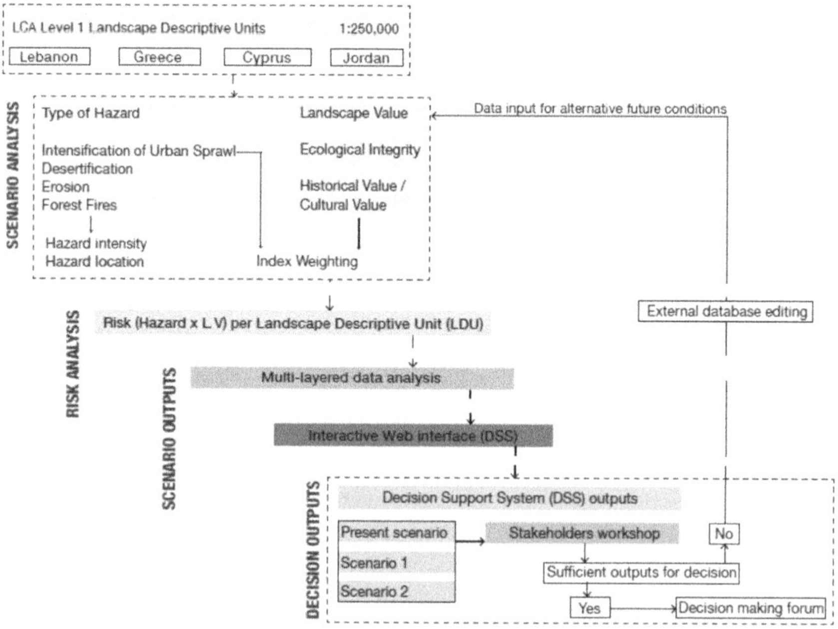

Figure 9. Methodology: from Risk Assessment to Decision Support System. LCA, Landscape Character Attribute; LV, Landscape Value.

The LDSS does not require the end user to have any knowledge or experience in LCA. It is embedded in a user-friendly GIS application where the LCA and LRA parameters cannot be edited by the users but only by experienced and well-trained personnel inside the ministries or regional and local planning offices.

The LDSS offers an integrative framework in landscape risk assessment and scenario building for the support of land-use decisions via a map-centered approach. It informs decision-makers about the risks affecting the land and thus will allow stakeholders to use real scenarios when determining new mitigation, conservation, restoration, or prevention measures. In environmental studies, the Intergovernmental Panel on Climate Change (IPCG) scenarios are described as 'images of the future, or alternative futures' that are neither predictions nor forecasts, but an alternative image of how the future might unfold [58].

MEDSCAPES Scenario Generator (SG) has been developed with the following capabilities:

- Credible scenarios are translated into changes on the GIS map.

- The SG queries the user for the appropriate information.
- The scenario(s) are then processed through the model.

The most important value of the LDSS lies in its capacity to easily communicate the results of the LCA and RA to a large and diverse audience through the scenarios' maps and for its interactivity which thus gives users the possibility to manipulate the data and spatially visualize the related scenarios. The most important function of the scenarios is that they act as crucial bridges between environmental/landscape science and policy. They influence policymaking by summarizing and synthesizing scientific knowledge in a form that can be used by policymakers to develop policies [55].

LDSS Functionality

The LDSS consists of a database, spatial analysis and bi-dimensional visualization of LCA and risk maps, Graphical User Interface (GUI), and a simulation model. The database supports cartographic info and permit the users to build spatial relations between coordinates and attributes. It contains landscape parameters that are related to the selected study areas for each of the four partner countries (i.e., Landforms, Soils, Land-cover, LCT, Erosion, Forest fire, Settlement, etc.), and coordinates for LDU polygons that allowed the partner countries to be segmented into the assessed LDUs. The spatial analysis and bi-dimensional visualization of risk maps enable the user to interactively explore two-dimensional LCA, hazards, landscape values, and risk maps in the selected areas of the four East Mediterranean partner countries. The GUI facilitates the interaction of the final user with the system. The simulation model permits the user to spatially create scenarios by rank modification or weight modification. The tool and the GUI closely follow the conceptual model of the LRA, enabling the user to perform the step by step assessment, varying input parameters, testing and comparing different combinations of scenarios.

The user interface is designed in a way to easily allow users to interact, create, and visualize through maps the result(s) of their choice(s).

4. Results

The above-described methodology was first tested on Lebanon's study area. This trial and error period clarified what kind of data would be specifically needed from the partner countries, and the different limitations faced. The concept of the model was then shared in phases of its development with the partners, where Skype meetings took place to share views on the model and the requested data. The final model reflects the difficulties encountered and the necessity to develop a system able to use and process such diversified information and to produce comparable results. Once the methodology had been built the expectation was to adopt it in the different countries and to further develop it to better respond and fit to the countries' planning rules and logic. In any case, it was a trial to consider the landscape, and the relevance that it is acquiring all over the world, with the definition of hazards and typology/aim of risk management.

4.1. Hazard Maps

Desertification, Erosion, Forest Fire, and Intensification of urban sprawl hazard maps were produced for all of Cyprus and for the selected area of Lebanon. In Greece and Jordan, Forest Fire was not assessed due the impossibility of getting the right information to be reworked. Furthermore, while the Desertification, Erosion, and Forest Fire hazards arise from different sources of stressors that were not assessed during the length of the MedScapes project, the final outcome is the result of data collected and compared using the available forms of national information that were reworked and qualitatively ranked due the uncertainty of the materials (Figures 10–13).

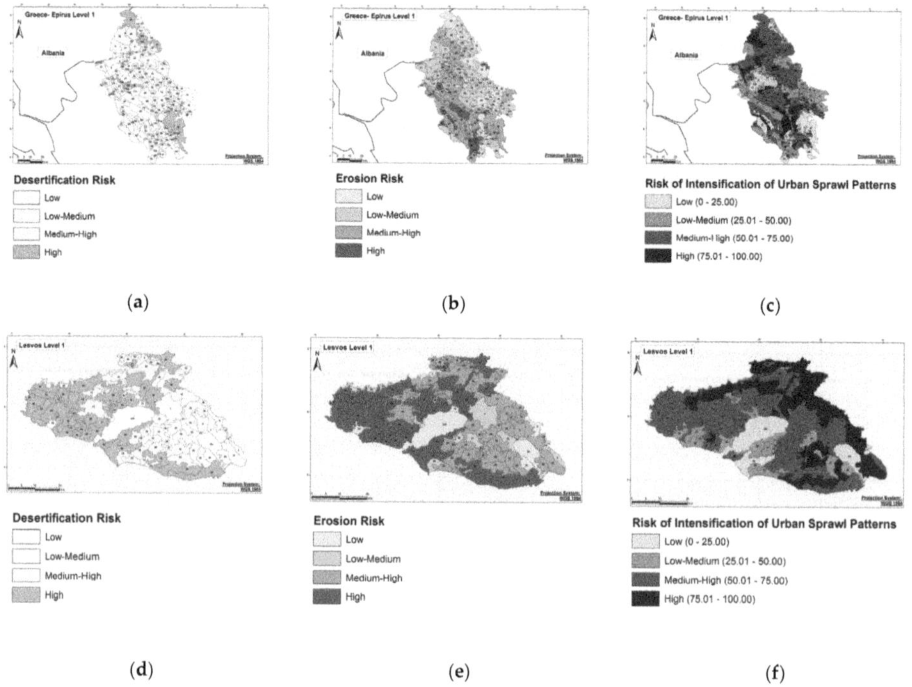

Figure 10. Hazard Maps for LDU's of Desertification (left), Erosion (middle), and Intensification of urban sprawl (right) in Greece: Epirus (**a**–**c**) and Lesvos (**d**–**f**).

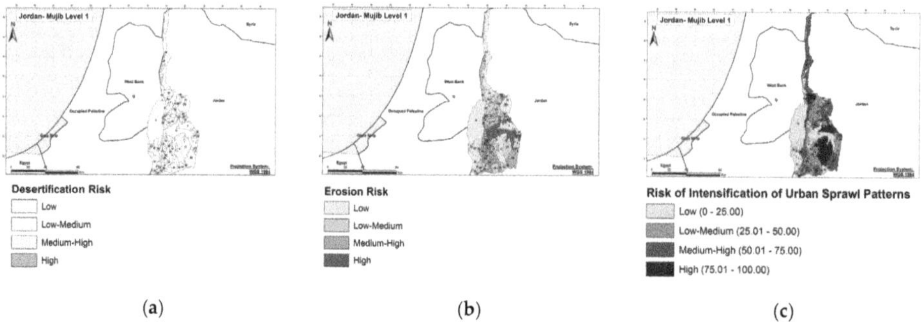

Figure 11. *Cont.*

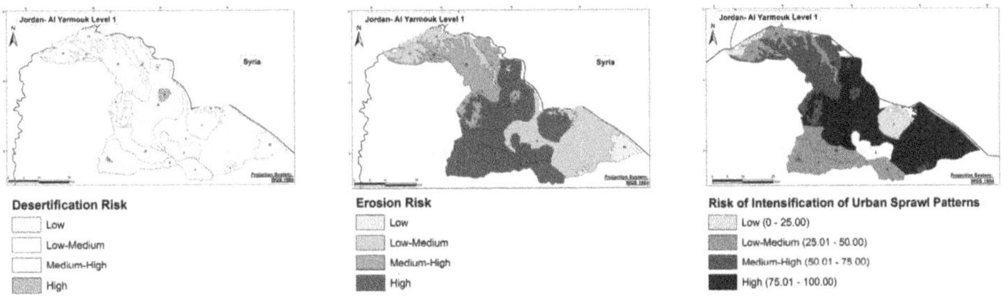

(d) (e) (f)

Figure 11. Hazard Maps for LDU's of Desertification (left), Erosion (middle), and Intensification of urban sprawl (right) in Jordan: Al-Yarmuk (**a**–**c**) and Mujib (**d**–**f**).

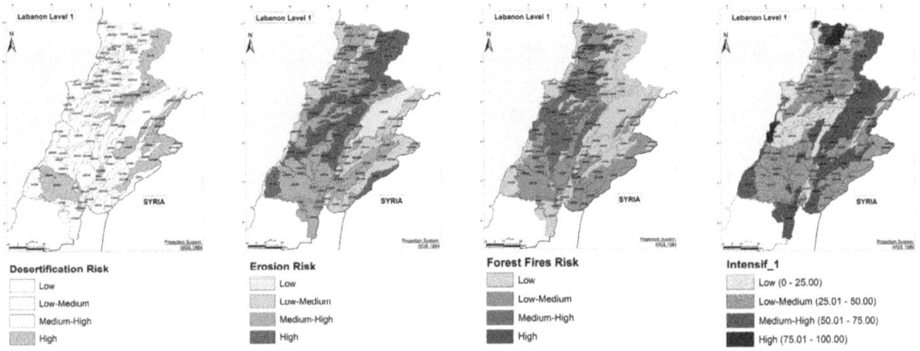

(a) (b) (c) (d)

Figure 12. Hazard Maps for LDU's of Desertification (**a**), Erosion (**b**), Forest Fire (**c**), and Intensification of urban sprawl (**d**) in Lebanon.

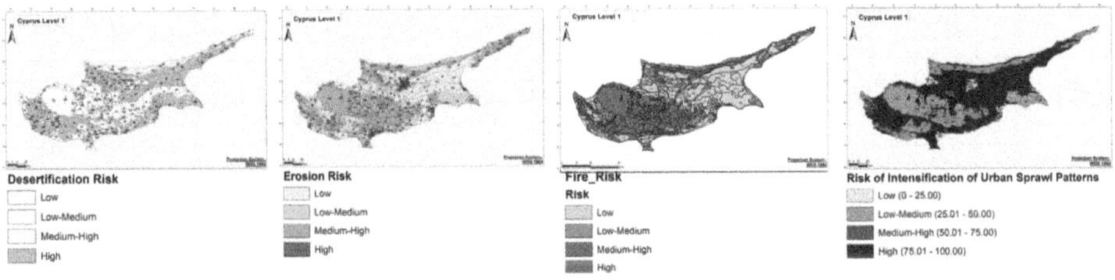

(a) (b) (c) (d)

Figure 13. Hazard Maps for LDU's of Desertification (**a**), Erosion (**b**), Forest Fire (**c**), and Intensification of urban sprawl (**d**) in Cyprus.

4.2. Landscape Value Maps

At the end of the Landscape Value assessment we produced a LV map for all the selected areas for each partner's country (Figures 14–16).

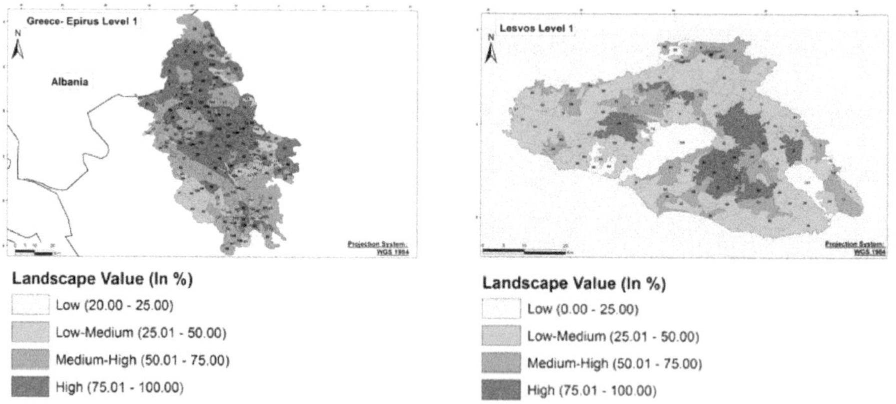

(**a**) (**b**)

Figure 14. Landscape Value (LV) maps in Greece: (**a**) Epirus and (**b**) Lesvos.

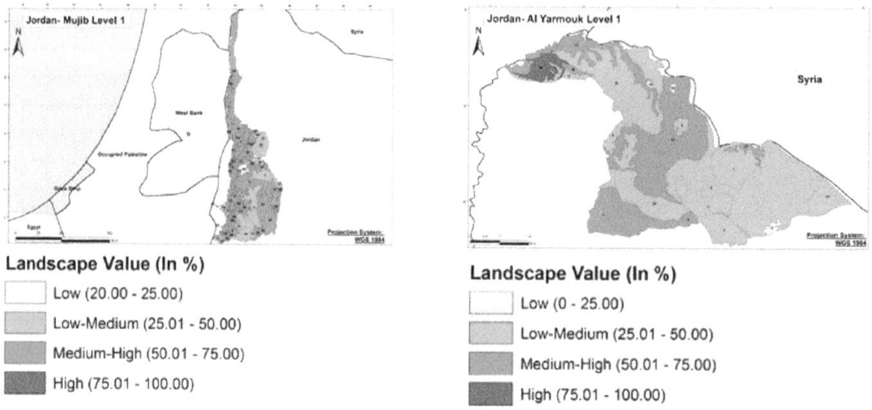

(**a**) (**b**)

Figure 15. Landscape Value (LV) maps in Jordan: (**a**) Mujib and (**b**) Al-Yarmuk.

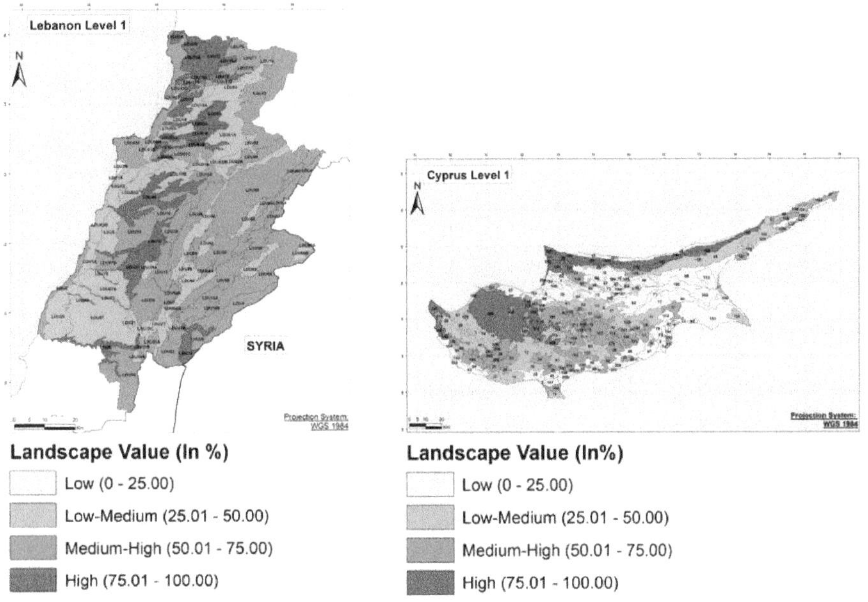

(a) (b)

Figure 16. Landscape Value (LV) maps in (**a**) Lebanon and (**b**) Cyprus.

4.3. Landscape Risk Maps

The overall landscape risk maps produced within the LRA-LDSS are the result of the MedScapes risk equation that integrates hazard versus landscape value for the LDUs in the four analyzed countries: Cyprus, Jordan, Greece, and Lebanon (Figures 17–19).

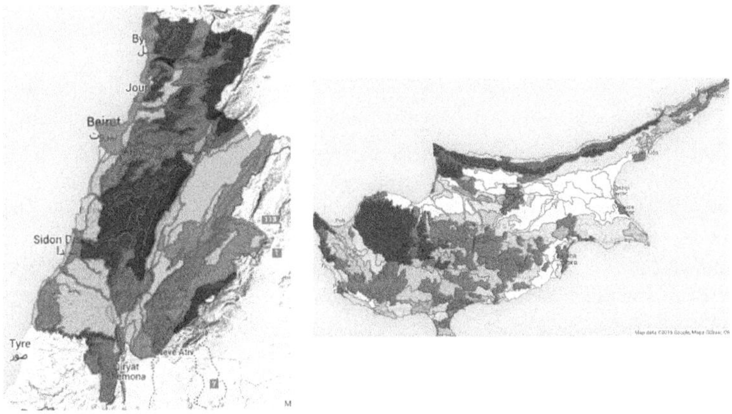

(a) (b)

Figure 17. Landscape Risk Assessment (LRA) maps for Lebanon (**a**) and Cyprus (**b**).

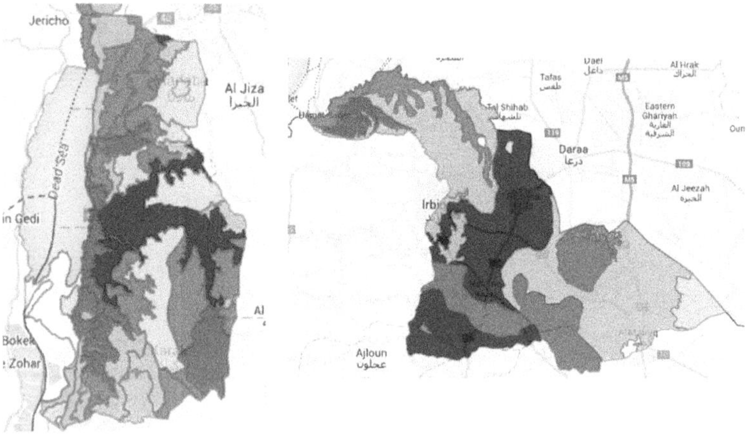

(a) (b)

Figure 18. Landscape Risk Assessment (LRA) maps for the two areas in Jordan: (**a**,**b**).

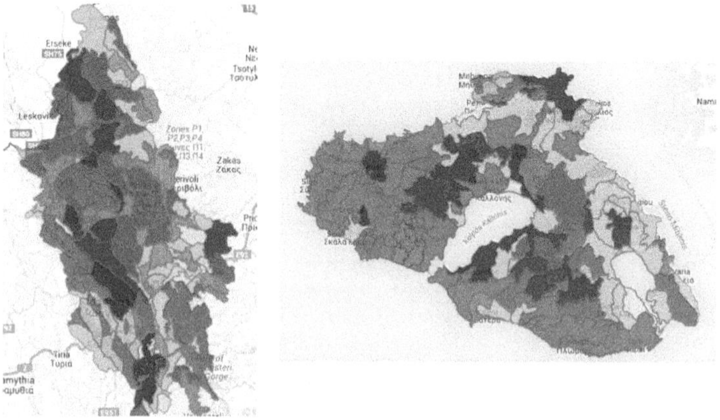

(a) (b)

Figure 19. Landscape Risk Assessment (LRA) maps for the two areas in Greece: (**a**,**b**).

The two-dimensional matrix for a risk legend enabled us to spatially represent the results of our study (Figure 20). Furthermore, it plays an important role in the evaluation process of the effects of the risk in relation to the loss of landscape character of the affected area. For each chosen Hazard in every selected area, it enables us to estimate and judge the risk of a reduction in LV that is directly proportional to the intensity of the hazard. In the upper part of matrix, highlighted as 'high risk of loss', the dark colors detect the LDU's that are subjected to a greater risk. It means that they are areas of particular interest for protection/conservation land-use decisions, and indicates a more pressing need of attention. In Lebanon, 57.29% of LDUs lie within the high risk of loss zone, 31% lie within the medium risk of loss zone, and 9.79% lie within the low risk of loss zone (Figure 21).

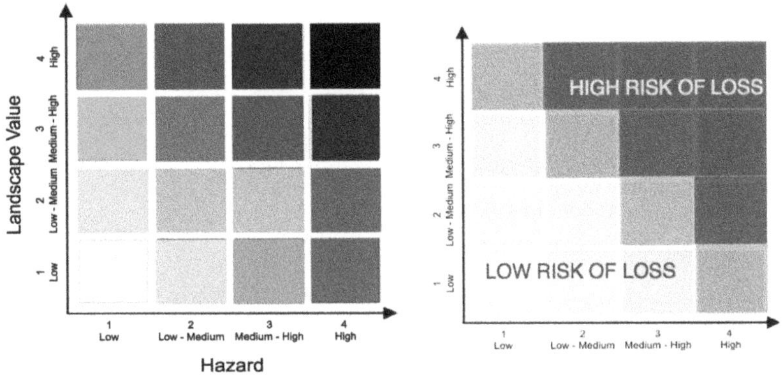

Figure 20. Landscape Risk Assessment (LRA) legend.

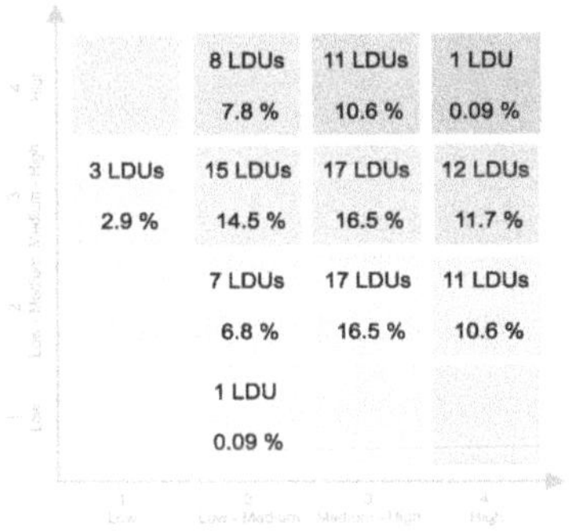

Figure 21. Number of LDUs per value of risk of loss of landscape value.

4.4. Landscape Decision Support System

The results of the spatial Risk Assessment are available online on the landscape RA and DSS interactive framework created during the MedScapes project. Using this tool, the users can visualize on GIS maps the outcome of the overall study and can generate spatially-based scenarios by altering the rank and weigh of the indicators. In the first case (Figure 22, the users can manipulate the Ecological Integrity components by changing the rank value of the Habitat Continuity, Dominant Habitat Type, and Naturalness and/or they can alter the Intensification of Urban Sprawl hazard components modifying the Refugees, Formation, Landform, and Settlement rank values. The operator can either start by filtering LDUs based on the indexing parameters, or by directly editing the LDUs of their choice, depending on their areas of interest. The higher the rank value assigned to the weighting of the parameter the more conducive it is to Landscape Value and Intensification of Urban Sprawl.

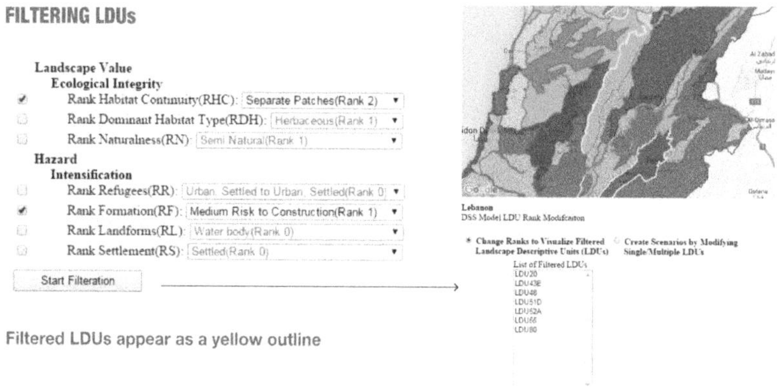

Figure 22. LDSS scenarios create by rank modification.

In the case of the scenarios generate by weight modification, the Landscape Values and Intensification of Urban Sprawl pressures could be changed bearing in mind that the total value of the weight percentage can't exceed the 100 percent (Figure 23). The scenarios generated by the model allow users to see a case series of possible physical transformations of the territory. Those changes, due to the modified intensity of the hazards and of the LV, will have repercussions on the character of the landscape that will be readable and evaluable with the help of the matrix. While identifying and ranking the LDU areas at greatest risk of loss of landscape character the scenarios will help decision-makers in choosing the needed measures and actions to reduce the exposure to the selected hazards.

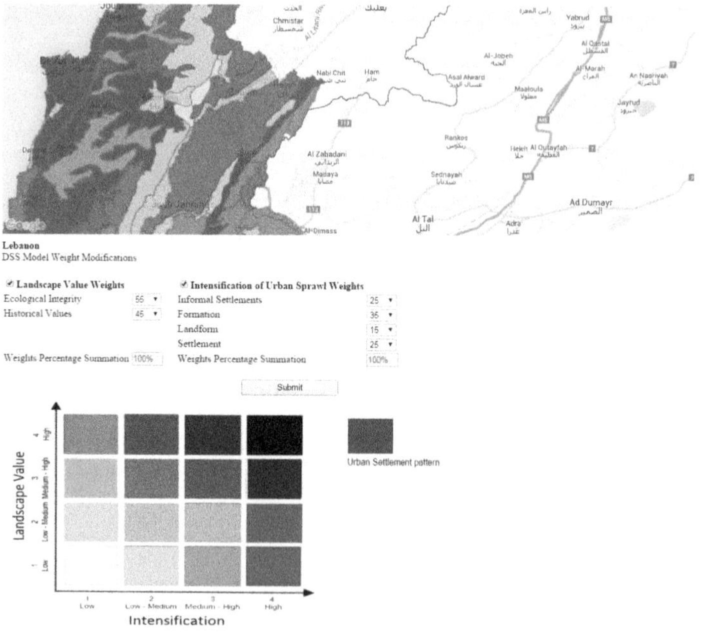

Figure 23. LDSS scenarios created by weigh modification.

5. Discussion

The LRA LDSS model conceived for MEDSCAPES aimed to develop a tool that helps in the process of landscape decision-making for the protection of the natural and cultural heritage of the Eastern Mediterranean area. It was formulated by building on the Landscape Characterization procedure by relating/integrating its findings into the Landscape risk assessment within the framework of a decision support tool. This was a means of assisting administrators/stakeholders in the planning process through scenarios that were drawing modifying weight and rank of the indicators. The literature on LCA, RA, and DSS reveals that in the past decades different models have been built to facilitate the decision process, thus providing a basis for the definition of adaptation measures and planning strategies. A study by Frede et al. (2002) discusses sustainable planning through understanding multiple functions of the landscape, based on spatial analyses and assessments [59]. The MULBO model offers a framework of tools to integrate assessments of landscape functions into land use decision-making [21].

Ryan et al. (2006) discuss opportunities to involve stakeholder participation in land use decision problems for greenway planning applications [60]. Most of the SDSS or C-DSS models contemplate the inclusion of stakeholder participation, i.e., the Collaborative Multi-Criteria Spatial Decision Support System (C-MC SDSS) for the evaluation of multifunctional landscape according to a human smart perspective [57]. In the C-MC SDSS, awareness and multidimensional reading of the landscape are considered the two important focuses for a smart local development. In the East Mediterranean, there is no evidence of effective public participation in the decision-making process. In Lebanon, for example, top-down "modernist" urban planning approaches—which continuously favor local elites such as politicians and landowners and pay little attention to social and community interests—are still commonly observed [61]. Citizens' involvement in the planning process is implemented in only a few scattered situations. Furthermore, stakeholder involvement in the LCA process during the MEDSCAPES project was not very successful due to the low rate of responses we were able to collect through the interviews we conducted during the first phase of the project. Thus, we were not able to incorporate the insufficient data gathered in the LDSS database, but we aimed at spreading awareness on landscape related topics by visualizing through our user's interface the effects that today's actions could produce in our territories in the East Mediterranean. Through workshop and seminars training we were able to introduce the audience to the interface and to let them visualize, for example, the consequences that the spread of urbanization could have on future development and, in particular, on the landscape seen as a human-ecologic system, which can offer a wide range of benefits with positive effects on the quality of life of the local community. On the other hand, our tool foresees the user's interaction as manipulation of rank and weight of indicators with particular emphasis on intensification of urban sprawl and ecological integrity. The user-friendly interface allows multiple users to provide input and generate real-time output to support negotiated spatial decisions [62]. The LDSS framework, through weighing and evaluating alternatives, will help decision makers in conceptualizing alternatives and thus in considering the related decisions for future planning. Once an alternative is selected and implemented the results could be compared with the produced scenarios and then a revaluation of the criteria could be done. The scenarios act as crucial bridges between environmental/landscape science and policy. They influence policymaking by summarizing and synthesizing scientific knowledge in a form that can be used by policymakers to develop policies [63]. Our method, which did not incorporate quantitative information on the public preferences and values attributed to the landscape, has the advantage of being easy to perform and it has the capacity to easily communicate the outcomes of the LCA and LRA to a large and diverse audience through the scenarios' maps; it also has the benefit of being interactive, thus giving to the users the possibility to manipulate the data and spatially visualize the related scenarios.

We believe it is necessary to underline that the LDSS is a first-level tool and a result of the research conducted during the length of the MEDSCAPES project. It was conceived in a way to be further upgraded by integrating up-to-date info in order to produce more credible scenarios and in order for the involved parties to be able to use it and to further adapt it to their need and regulations. The workshops and the training session between the MEDSCAPES developers and the interested end-users (regional and planning department, ministries, etc.) allowed us to test the tool and to identify problems and options in order to ameliorate and calibrate the system to a more context-based situation. In Lebanon, for instance, the LDSS is in the process of revision and development under the Food and Agriculture Organization of the United Nations (FAO) grant 'Forest and Landscape Restoration Mechanism'. Thus, it has been further adjusted in order to better fit the research objective and to assess, and map the loss of landscape character in relation to the forested LDUs, and to develop scenarios of the different probabilities of landscape character loss/change under the threats of different hazards. Like other methods, such as DESYCO [6], our tool could be applied in different geographical contexts, and spatial scales, however, it does not overcome the issue of uncertainty assessment toward some of the indicators and it does not allow an evaluation of social risk perception.

6. Conclusions

This study, conducted at the regional scale of the Eastern Mediterranean basin, is an important attempt to assess and collate the landscape character of the region by individuating and using comparable indicators to evaluate the risk of loss of the landscape value of the East Mediterranean region. Furthermore, the LDSS enables users and stakeholders to build helpful scenarios in the planning process of the selected areas. The generation and selection of scenarios advocate for a bottom-down approach to landscape planning that could complement the conventional practice, and it constitutes a step forward into sustainable land use protection and management. Many authors have supported the use of risk assessment and decision analysis in the selection of implementable solutions to environmental problems [64], however, only a small number of studies integrate different landscape features, such as land-use, vegetation, geology, hydrography, cultural and archaeological elements into the assessment methods [24]. Our model was successful in incorporating and working ecological and cultural features of the East Mediterranean landscape but did not take into consideration the socio-economical and human health components. The reason lays in the objectives of the research and in the typology of landscape assessment the partners of the project agreed on using to conduct our study. Results show the importance of landscape protection policies in addressing the main threats to landscape values with particular regards towards the urban sprawl processes whose consequences on the landscape as a whole are of great magnitude.

The methodology was designed using previously tested research studies that, following a comparison to our objectives, were reworked and adapted to better suit the aim of our study. It focuses mainly on the physical components of the landscape that were evaluated under the landscape ecology lens with the incorporation of few cultural assets. Though limited, these assets were important in portraying a more thorough picture of the Mediterranean landscape character and of the threats to which it is subjected.

The opportunities presented by a landscape-based approach for the European continent mark out the landscape as a possible new paradigm for a local development model, with the aim of harmonious integration of social, economic, and environmental factors in space and time [65]. It represented an important step in the individuation of a method and related tool that is able to assess and compare the Landscape Risk at the scale of the Eastern Mediterranean basin. Furthermore, the methodology and framework developed within the MedScapes project are playing an important and strategic role in advocating the importance of the landscape in the planning process, and in spreading awareness on the protection and management of natural and cultural heritage in the Eastern Mediterranean region.

The designed LRA and LDSS tools require further development that could allow the system to incorporate the assessment of socio and economic factors related to the landscape character of the East Mediterranean Region in general and to the selected countries in particular. This inclusion will portray more holistic and inclusive scenarios, thus allowing the possibility for the monetary consequence of the assessed landscape loss and its repercussions on social wellbeing to be estimated and quantified.

Acknowledgments: The research was conducted with the financial support of the cross-border cooperation 'Mediterranean Sea Basin Programme-ENPI' of the European Union. The grant is not supporting the costs to publish in open access. Maria Gabriella Trovato wants to express her gratitude to her University, AUB, and her department in particular, LDEM, for the support and help throughout the duration of the project. A special thanks to all the governmental institutions that directly and indirectly contributed to the smooth development of the research by giving access to resources and data necessary to run our project.

Author Contributions: Maria Gabriella Trovato and Dana Ali conceived and designed the experiments; Jessica Nicolas and Ammar El Halabi performed the experiments; Jessica Nicolas, Dana Ali, and Ammar El Halabi analyzed the data; Maria Gabriella Trovato, Dana Ali, and Sarah Meouche wrote the paper.

Conflicts of Interest: The authors declare no conflict of interest. The founding sponsors had no role in the design of the study; in the collection, analyses, or interpretation of data; in the writing of the manuscript, and in the decision to publish the results.

References

1. Nogué, J.; Wilbrand, S.M. Landscape identities in Catalonia. *Landsc. Res.* **2017**, 1–12. [CrossRef]
2. Secretary of State for Foreign and Commonwealth Affairs. *European Landscape Convention, Florence*; Treaty Section, Foreign and Commonwealth Office: London, UK, 2000.
3. Kapustka, L.; Galbraith, H.; Luxon, M.; Yocum, J. Using landscape ecology to focus ecological risk assessment and guide risk management decision-making. *Toxicol. Ind. Health* **2001**, *17*, 236–246. [CrossRef] [PubMed]
4. Kochanek, K.; Tynan, S. The environmental risk assessment for decision support system for water management in the vicinity of open cast mines (ds wmvoc). *Ukio Technol. Ekon. Vystym.* **2010**, *16*, 414–431. [CrossRef]
5. ARMONIA Project. Report on the European scenario of technological and scientific standards reached in spatial planning versus natural risk management. In *The Adaptiveness of IWRM: Analysing European IWRM Research*; Timmerman, J., Pahl-Wostl, C., Moltgen, J., Eds.; IWA: London, UK, 2008.
6. Torresan, S.; Critto, A.; Rizzi, J.; Zabeo, A.; Furlan, E.; Marcomini, A. DESYCO: A decision support system for the regional risk assessment of climate change impacts in coastal zones. *Ocean Coast. Manag.* **2016**, *120*, 49–63. [CrossRef]
7. Agostini, P.; Pizzol, L.; Critto, A.; D'Alessandro, M.; Zabeo, A.; Marcomini, A. Regional risk assessment for contaminated sites Part 3: Spatial decision support system. *Environ. Int.* **2012**, *48*, 121–132. [CrossRef] [PubMed]
8. Brown, G.; Braby, L. An analysis of the relationships between multiple values and physical landscapes at a regional scale using public participation GIS and landscape character classification. *Landsc. Urban Plan.* **2012**, *107*, 317–331. [CrossRef]
9. Beverly, J.; Uto, K.; Wilkes, J.; Bothwell, P. Assessing spatial attributes of forest landscape values: An internet-based participatory mapping approach. *Can. J. For. Resour.* **2008**, *38*, 289–303. [CrossRef]
10. Brown, G.; Reed, P. Validation of a forest values typology for use in national forest planning. *For. Sci.* **2000**, *46*, 240–247.
11. Clement, J.M.; Cheng, A.S. Using analyses of public value orientations, attitudes and preferences to inform national forest planning in Colorado and Wyoming. *Appl. Geogr.* **2010**, *31*, 393–400. [CrossRef]
12. Brown, G.; Weber, D. Measuring change in place values using public participation GIS (PPGIS). *Appl. Geogr.* **2012**, *34*, 316–324. [CrossRef]
13. Tyrväinnen, L.; Mäkinen, K.; Schipperijn, J. Tools for mapping social values of urban woodlands and other green areas. *Landsc. Urban Plan.* **2007**, *79*, 5–19. [CrossRef]
14. Brown, G. Mapping landscape values and development preferences: A method for tourism and residential development planning. *Int. J. Tour. Res.* **2006**, *8*, 101–113. [CrossRef]
15. Raymond, C.; Brown, G. A spatial method for assessing resident and visitor attitudes toward tourism growth and development. *J. Sustain. Tour.* **2007**, *15*, 520–540. [CrossRef]
16. Alessa, N.; Kliskey, A.; Brown, G. Social–ecological hotspots mapping: A spatial approach for identifying coupled social-ecological space. *Landsc. Urban Plan.* **2008**, *85*, 27–39. [CrossRef]
17. Nielsen-Pincus, M. Land Use, Resource Management, and Place: A Case Study in Three Counties of Northern Idaho and Northeastern Oregon. Ph.D. Thesis, University of Idaho, Moscow, ID, USA, 2007.
18. Pocewicz, A.; Schnitzer, R.; Nielsen-Pincus, M. *The Social Geography of Southern Wyoming: Important Places, Development, and Natural Resource Management*; The Nature Conservancy: Lander, WY, USA, 2010; 16p.

19. Raymond, C.; Brown, G. Assessing spatial associations between perceptions of landscape value and climate change risk for use in climate change planning. *Clim. Chang.* **2011**, *104*, 653–678. [CrossRef]
20. La Rosa, D. The observed landscape: Map of visible landscape values in the province of Enna (Italy). *J. Maps* **2011**, *7*, 291–303. [CrossRef]
21. Meyer, B.C.; Grabaum, R. MULBO: Model framework for multicriteria landscape assessment and optimization. A support system for spatial land use decision. *Landsc. Res.* **2008**, *33*, 155–179. [CrossRef]
22. Clay, G.R.; Daniel, T.C. Scenic landscape assessment: The effects of land management jurisdiction on public perception of scenic beauty. *Landsc. Urban Plan.* **2000**, *49*, 1–13. [CrossRef]
23. Meyer, B.C. Landscape assessment. In *Landscape Balance and Landscape Assessment*; Krönert, R., Steinhardt, U., Volk, M., Eds.; Springer: Berlin/Heidelberg, Germany, 2001; pp. 203–250.
24. La Rosa, D.; Martinico, F. Assessment of hazards and risks for landscape protection planning in Sicily. *J. Environ. Manag.* **2013**, *127*, S155–S167. [CrossRef] [PubMed]
25. Antrop, M. Sustainable landscapes: Contraction, fiction or utopia. *Landsc. Urban Plan.* **2006**, *75*, 187–197. [CrossRef]
26. Bacon, C.M. Disaster Risk and Sustainable Development. In *The Routledge Handbook of Hazards and Disaster Risk Reduction*; Wisner, B., Gaillard, J.C., Kelman, I., Eds.; Routledge: London, UK, 2011; pp. 156–167.
27. Dransch, D.; Etter, J.; Walz, U. Maps for natural risk management. In Proceedings of the 22nd International Cartographic Conference, La Coruna, Spain, 9–16 July 2005.
28. Menoni, S.; Margottini, C. *Inside Risk: A Strategy for Sustainable Risk Mitigation*; Foreword; Springer: Milan, Italy, 2011.
29. Walker, G.; Deeming, H.; Margottini, C.; Menoni, S. Introduction to Sustainable Risk Mitigation for a more resilient Europe. In *Inside Risk: A Strategy for Sustainable Risk Mitigation*; Foreword; Menoni, S., Margottini, C., Eds.; Springer: Milan, Italy, 2011.
30. Martinico, F.; La Rosa, S.D. The Use of GIS in Landscape Protection Plan in Sicily. In *Urban and Regional Data Management (UDMS) Annual 2009*; Krek, A., Rumor, M., Zlatanova, S., Fendel, E., Eds.; Taylor & Francis: London, UK, 2009; pp. 315–325.
31. De Ayala, A.; Hoyos, D.; Mariel, P. Suitability of discrete choice experiments for landscape management under the European Landscape Convention. *J. For. Econ.* **2015**, *21*, 79–96. [CrossRef]
32. Johnston, R.J.; Duke, J.M. Willingness to pay for agricultural land preservation and policy process attributes: Does the method matter? *Am. J. Agric. Econ.* **2007**, *89*, 1098–1115. [CrossRef]
33. Burley, J.B.; Burley, C.J. A risk assessment of landscape hazards for building sites in the Front Range Mountains of Colorado. *Landsc. Res.* **1996**, *21*, 137–158. [CrossRef]
34. Fernandes, J.P. Landscape ecology and conservation management e evaluation of alternatives in a highway EIA process. *Environ. Impact Assess. Rev.* **2000**, *20*, 665–680. [CrossRef]
35. Antrop, M. Background concepts for integrated landscape analysis. *Agric. Ecosyst. Environ.* **2000**, *77*, 17–28. [CrossRef]
36. Williamson, D. Scenic perceptions of Australian landscapes. *Landsc. Aust.* **1979**, *2*, 94–101.
37. Leonard, M.; Hammond, R. *Landscape Character Types of Victoria-With Frames of Reference for Scenic Quality Assessment*; Forest Commission of Victoria: Melbourne, Australia, 1984.
38. Daniel, T.C.; Vining, J. Methodological Issues in the assessment of landscape quality. In *Behaviour and the Natural Environment*; Altman, I., Wohwill, J.F., Eds.; Plenum Press: New York, NY, USA, 1983; pp. 39–83.
39. Regional Forest Agreements Act 2002. Australia.
40. Butler, A. Dynamics of integrating landscape values in landscape character assessment: The hidden dominance of the objective outsider. *Landsc. Res.* **2016**, *41*, 239–252. [CrossRef]
41. Brown, G. Mapping spatial attributes in survey research for natural resource management: Methods and applications. *Soc. Nat. Resour.* **2004**, *18*, 17–39. [CrossRef]
42. Brown, G.; Raymond, C. The relationship between place attachment and landscape values: Toward mapping place attachment. *Appl. Geogr.* **2007**, *27*, 89–111. [CrossRef]
43. Brown, G.; Raymond, C.M. Methods for identifying land use conflict potential using participatory mapping. *Landsc. Urban Plan.* **2014**, *122*, 196–208. [CrossRef]
44. Jones, M. Analysing landscape values expressed in planning conflicts over change in the landscape. In *Re-Marc-able Landscapes: Marc-ante Landschappen*; Eetvelde, V., Van Sevenant, M., Van De Velde, L., Eds.; Academia Press: Gent, Belgium, 2009; pp. 193–205.

45. Stephenson, J. The cultural values model: An integrated approach to values in landscapes. *Landsc. Urban Plan.* **2008**, *84*, 127–139. [CrossRef]
46. Henningsson, M.; Blicharska, M.; Antonson, H.; Mikusiński, G.; Göransson, G.; Angelstam, P.; Jönsson, S. Perceived landscape values and public participation in a road-planning process—A case study in Sweden. *J. Environ. Plan. Manag.* **2014**, *58*, 631–653. [CrossRef]
47. Hay, R. A rooted sense of place in cross-cultural perspective. *Can. Geogr.* **1998**, *42*, 245–266. [CrossRef]
48. Gray, J. A rural sense of place: Intimate experience in planning a countryside for life. *Plan. Theory Pract.* **2013**, *4*, 93–96. [CrossRef]
49. De Angelis, D.L.; Pearlstine, L.; Mazzotti, F.J.; Barnes, T.; Duever, M.; Starners, J. *Spatial Decision Support Systems for Landscape Ecological Evaluations in the Southwest Florida Feasibility Study*; US Geological Survey Report; U.S. Department of the Interior: Washington, DC, USA, 2004.
50. Greiving, S.; Fleischhauer, M.; Lückenkötter, J. A methodology for an integrated risk assessment of spatially relevant hazards. *J. Environ. Plan. Manag.* **2006**, *49*, 1–19. [CrossRef]
51. Rauscher, H.M. Ecosystem management decision support for Federal forests in the United States: A review. *For. Ecol.* **1999**, *114*, 173–197. [CrossRef]
52. He, J.; Sun, Y. Applying a spatial decision support system to the integrated regional planning of China. *Environ. Plan. B Urban Anal. City Sci.* **2015**, *42*, 1161–1176. [CrossRef]
53. Fedra, K.; Feoli, E. GIS technology and spatial analysis in coastal zone management. *EEZ Technol.* **1998**, *3*, 171–179.
54. Wilkerson, G.W.; McAnally, W.H.; Martin, J.L.; Ballweber, J.A.; Pevey, K.C.; Diaz-Ramirez, J.; Moore, A. Latis: A spatial decision support system to assess low-impact site development strategies. *Adv. Civ. Eng.* **2010**. [CrossRef]
55. Simoes, M. GIS and spatial decision support system for environmental degradation monitoring. In Proceedings of the 7th International Seminar on GIS in Developing Countries, Johor, Malaysia, 10–12 May 2004.
56. Nakicenovic, N.; Alcamo, J.; Davis, G.; de Vries, H.J.M.; Gaffin, S.; Gregory, K.; Grubler, A.; Jung, T.Y.; Kram, T.; Lebre La Rovere, E.; et al. *Intergovernmental Panel on Climate Change (IPCG) Special Report on Emission Scenarios (SRES)*; Intergovernmental Panel on Climate Change (IPCC): Geneva, Switzerland, 2000.
57. Attardi, R.; Cerreta, M.; Poli, G. A Collaborative Multi-Criteria Spatial Decision Support System for Multifunctional Landscape Evaluation. In *Computational Science and Its Applications—ICCSA 2015*; Lecture Notes in Computer Science; Gervasi, O., Murgante, B., Misra, S., Gavrilova, M.L., Rocha, A.M.A.C., Torre, C.M., Taniar, D., Apduhan, B.O., Eds.; Springer: Cham, Switzerland, 2015.
58. Alcamo, J. *Scenarios as Tools for International Environmental Assessments*; Environmental Issue Report; European Environment Agency (EEA): Copenhagen, Denmark, 2001.
59. Frede, H.-G.; Bach, M.; Fohrer, N.; Möller, D.; Steiner, N. Multifunktionalitat der Landschaft—Methoden und Modelle. *Petermanns Geogr. Mitt.* **2002**, *146*, 58–63.
60. Ryan, R.L.; Fábos, J.G.; Allan, J.J. Understanding opportunities and challenges for collaborative greenway planning in New England. *Landsc. Urban Plan.* **2006**, *76*, 172–191. [CrossRef]
61. El Asmar, E.A.; Ebohonb, J.O.; Taki, A. Bottom-up approach to sustainable urban development in Lebanon: The case of Zouk Mosbeh. *Sustain. Cities Soc.* **2012**, *2*, 37–44. [CrossRef]
62. Arciniegas, G.; Janssen, R. Spatial decision support for collaborative land use planning workshops. *Landsc. Urban Plan.* **2012**, *107*, 332–342. [CrossRef]
63. Urban and Regional Data Management. *UDMS Annual 2009*; Krek, A., Rumor, M., Zlatanova, S., Fendel, E.M., Eds.; Taylor & Francis: London, UK, 2009.
64. Bonanno, E.J.; Apostolakis, G.E.; Salter, P.F.; Ghassemi, A.; Jennings, S. Application of risk assessment and decision analysis to the evaluation, ranking and selection of environmental remediation alternatives. *J. Hazard. Mater.* **2000**, *71*, 35–57. [CrossRef]
65. Agnoletti, M. Rural landscape, nature conservation and culture: Some notes on research trends and management approaches from a (southern) European perspective. *Landsc. Urban Plan.* **2004**, *126*, 66–73. [CrossRef]

MDPI
St. Alban-Anlage 66
4052 Basel
Switzerland
Tel. +41 61 683 77 34
Fax +41 61 302 89 18
www.mdpi.com

Land Editorial Office
E-mail: land@mdpi.com
www.mdpi.com/journal/land

Lightning Source UK Ltd.
Milton Keynes UK
UKHW021020100622
404228UK00003B/27

9 783039 217748